THE HEALTHY WORKPLACE

THE HEALTHY WORKPLACE

A BLUEPRINT FOR CORPORATE ACTION

WILLIAM M. KIZER

Chairman and Chief Executive Officer
Central States Health & Life Co. of Omaha
Omaha, Nebraska
Founder, Wellness Council of the Midlands
Omaha, Nebraska

JOHN WILEY & SONS
New York • Chichester • Brisbane • Toronto • Singapore

Library of Congress Cataloging-in-Publication Data:
Kizer, William M.
  The healthy workplace.

  (A Wiley medical publication)
  Bibliography: p.
  Includes index.
  1. Occupational health services. 2. Employee
assistance programs. 3. Industrial hygiene. I. Title.
II. Series. [DNLM: 1. Health Promotion—United States.
2. Occupational Health Services—United States.
3. Preventive Health Services—United States.
WA 412 K62h]
RC968.K58   1986   658.3′82   86-19095
ISBN 0-471-84531-0

Printed in the United States of America

10 9 8 7 6 5 4 3 2 1

To my wife, Lois,
a truly well human being

CONTRIBUTORS

Harold S. Kahler, Jr.
Executive Director
Wellness Council of the Midlands
Omaha, Nebraska

Fred W. Schott
Vice President, Training and Development
Central States Health & Life Co. of Omaha
Omaha, Nebraska
Past President and Board Member
Wellness Council of the Midlands
Omaha, Nebraska

FOREWORD

Business leaders increasingly are realizing that helping to keep employees healthy is an absolute cornerstone to health care cost containment.

Until recently, some executives were understandably skeptical about the feasibility of cost-effective worksite wellness systems. Such skepticism is fading quickly in light of new scientific data.

At Johnson & Johnson, for example, we developed the LIVE FOR LIFE Program in the late 1970s to serve as a comprehensive management system to improve employee health and contain health care costs. Scientific study results now show definite employee health and cost benefits. Employees offered the LIVE FOR LIFE Program for two and a half years or more incurred 40 percent lower hospitalization costs than employees who were not offered this worksite wellness program. Absenteeism was 18 percent less. Not surprisingly, these financial benefits were preceded by widespread health changes, including a dramatic reduction in smoking and a substantial increase in regular exercise.

It is our belief, therefore, that over time this type of program will be paid for out of reductions in health care costs.

Managers also are becoming more aware of the tremendous impact programs such as LIVE FOR LIFE can have on employee morale and productivity. Having the healthiest work force possible—with people highly committed to our organization—is just one of the many benefits provided by the LIVE FOR LIFE Program. We are convinced that its value to our company and our employees is beyond question.

The Healthy Workplace: A Blueprint for Corporate Action conveys to the reader—whether head of a small company or large corporation—exactly why worksite wellness is vital and how to implement a program, no matter how modest. A practical guide, the book clearly communicates specifically how management can benefit both the company and community.

Having initiated the first Wellness Council and served as chairman of the Health Insurance Association of America's Health Education Committee, Bill Kizer, perhaps more than any other individual, is the person to write the book on health programs in business.

Health care most certainly is in the midst of a revolution, and one of the most salutary changes is the increasing concern on the part of the public about staying well. I'm convinced that America's business leaders, once armed with current data, will be ready to help take advantage of these new insights.

James E. Burke
Chairman and Chief Executive Officer
Johnson & Johnson
New Brunswick, New Jersey

IS AMERICAN BUSINESS READY TO MAKE A DIFFERENCE? REMARKS FROM RONALD REAGAN

Disease prevention and health promotion are in all of our interests, not only for people in the insurance business, but for employees and employers throughout the wide spectrum of American enterprise. I don't need to tell you employers that the illness of your employees is a costly proposition. A healthier work force means higher productivity, reduced absenteeism, and less overtime. In the long run, it also means a reduction in the cost for employee health benefits.

Today we've conquered the old killers like smallpox, diphtheria, and polio. We understand that how each of us chooses to live will, more than anything else, determine our health.

Executives are in a position to provide leadership in this area, because working people spend about half their waking hours at work. With little or no financial investment, the employer can influence his or her employees to change some bad habits that heavily affect one's health.

Cigarette smoking is, perhaps, the best example. We all know how harmful it is. Well, the illness resulting from smoking is costly to both the smoker and his or her boss. A helping hand to assist employees to break the habit might be a wise investment.

Good eating and exercise habits are other areas where employers could use their influence. We are all aware of the fitness programs in Japanese companies. It's something you might think about.

The question now is whether you're willing to take the steps necessary to make a difference. Those of you who take this to heart have my sincere thanks.

America can only be as strong and healthy as its people, and, as in all things, the only lasting change that takes place comes when each of us does his part to make our country the good and decent place we want it to be.

Ronald Reagan
President of the United States
Remarks from the White House
Video Teleconference, March 1984

PREFACE

WORKSITE WELLNESS

- The latest management fad or destined to take its place as a pillar of American business practice?
- An expensive bit of unnecessary fluff or a solid investment that delivers two, five, and even seven to one for every dollar spent?
- A cheap public relations gimmick or evidence of genuine corporate concern for the quality of American life?
- Simple-minded slogans pushed on apathetic employees or one of the best vehicles ever designed for boosting morale, improving productivity, and building loyalty among employees?
- A cruel hoax thrown like a bone to organized labor as "real" health benefits or a genuine partnership of labor and management that finally pays attention to the preventive side of health care?
- A small ripple in this country's $425 billion annual health care ocean or the one thing that offers the most promise of having a significant long-range impact on the rising cost of health care?

Because it is so new and at the same time literally exploding throughout American business and industry, worksite wellness is a mixed bag of all of the above. But when done right—and that's what *The Healthy Workplace: A Blueprint for Corporate Action* will tell you—worksite wellness can deliver on the best things that it says about itself.

A healthier, more aware, better-educated American public, without doubt, has the greatest potential for lowering the costs of health care.

The shift from treatment to prevention has already created one shock wave after another in health care and its related industries.

Worksite wellness fits with both traditional and recent American management theories. It builds on the common theme that people have always been and will remain our greatest resource. The evidence that well-planned worksite wellness programs have a positive impact on morale, absenteeism, turnover, and productivity is already accumulating.

Worksite wellness promises to be so much more than good public relations. It is a unifying force. Who can argue with helping millions of Americans make better decisions about their health and take personal responsibility for their own well-being? It is clear that in worksite wellness what is good for the company is good for the worker, the worker's family, the community, and the country.

These themes and their promise are thoroughly developed in this book. This book explores what you can do as a business leader for your company and your community. If you are the chief executive officer (CEO) of IBM, Xerox, Control Data, Tenneco, General Motors, Metropolitan Life, or any of the other large corporations that already have worksite wellness programs underway, this book may not be as directly applicable to you as it is for the other "Fortune 4,500,000" companies. Or if you are one of the 65 million employees who work in firms employing 1,000 or fewer people—and you want to do something about health promotion where you work—this book is for you to read and bring to the attention of the decision maker.

But most Americans work in companies that employ fewer than 10 people. If you run such a company, congratulations! You're American business at its best, and this book has much to offer you. Although health promotion may be big business, it's not intended just for big business anymore.

In addition, the book should be useful to human resource and benefits personnel, fitness directors, insurance companies, health care practitioners, and hospital administrators.

A health magazine editor, T George Harris, bemoaned the fact that the idea of wellness at work is "still so new to American corporations that there is no single corporate executive who knows how to put it all together yet" (*Newsweek*, November 1984).

Here—finally—we have put it all together. This is the guide. This is the how-to for the corporate executive who gets excited about wellness and promoting health for employees but simply doesn't know where to begin or what to do. It's also for the skeptical corporate executive who has heard about worksite wellness but wants to know what it will do for the bottom line.

Health is not something you can buy for yourself or for your employees. Nor can doctors cure everything with a pill or an operation. Rene

Dubos, a distinguished biomedical scientist, stated: "To ward off disease or recover health, men as a rule find it easier to depend on the healers than to attempt the more difficult task of living wisely."

Living wisely, then, leads to wellness, no matter how you define it. One of the best ways to describe wellness is this:

> Wellness is a process of being aware of and of altering behavior toward a more successful physical, mental, emotional, psychological, occupational, and spiritual existence.

Wellness, as a process, can coexist with some strange bedfellows. For example, a terminally ill person can exhibit high levels of wellness with positive mental attitudes and a will to be as strong as possible, no matter how poor his or her physical health may be. On the other hand, a fit person who is the picture of health may be a child abuser. That person may be physically healthy, but he or she is certainly not achieving wellness.

Part I of *The Healthy Workplace* answers the tough questions that business leaders ask about worksite wellness. We talk about bottom-line payoffs to wellness in the language of business: how to make a return on an investment in wellness, how to minimize your risks, a look at the legal aspects and the tax implications, and a discussion of how wellness prepares the work force for the ups and downs of doing business in America.

The working section of this book is located in chapter 2, "A Business Plan for Health Promotion at Work." There you'll find the practical guide for business leaders and anyone else who is serving on a wellness committee, who has volunteered or been appointed to "check wellness out," or who wants to start something at work and just doesn't know where to begin or how to expand existing programs. The business plan takes you through the steps every corporate leader goes through when making any other business decision. Some of these suggestions may even revive half-hearted attempts to bring wellness into your workplace.

Working through the simple business plan outlined here will reveal that wellness programs are not just fitness rooms, exercise bicycles, and showers. Wellness targets many other lifestyle practices such as smoking, alcohol use, and nutrition. When I speak to industry leaders, I tell them that any company can get into wellness using the SANE approach to a healthy company: S for smoking, A for alcohol, N for nutrition, and E for exercise. And these are just the start. Other more complex and costly wellness programs grow out of these four areas. But the best part about the ideas presented in the SANE approach to a healthy company is that they can be done at no cost or low cost. Chapter 3 explains the SANE approach to a healthy company in detail.

Three of my business colleagues and I founded the Wellness Council of the Midlands (WELCOM) in 1982. WELCOM holds the distinction of being the first Wellness Council anywhere and the blueprint for other councils. The Wellness Council acts as a clearinghouse—a support network for large and small companies in all areas of business and industry and in all stages of planning and programming for wellness at work. Member companies send delegates to meetings to share ideas and exchange information. Since WELCOM boasts of having over 112 companies as members in America's heartland of Omaha, Nebraska, we invite you to learn about wellness from WELCOM. We want to show you how we promote wellness in our worksites and in our community.

The chicken-or-the-egg principle applies to worksite wellness programs and Wellness Councils. Which comes first? Does a company start a wellness program at work and then join other businesses in forming a Wellness Council, or does a company join a Wellness Council and draw on the council's resources to help design wellness programs for the company? Many companies have tackled both at the same time, each giving impetus to the other. The best answer to that question is whatever works. If you are fortunate enough to be in a community in which a Wellness Council is in early stages of forming, you're lucky. But for thousands of companies in hundreds of other communities, Wellness Councils do not yet exist.

On the flip side, many companies have been practicing their own brand of wellness for years. Some may have smoking policies in place but nothing else. Others may already have on-site exercise facilities or organized sports teams. These companies are ripe to take the lead in forming and to become charter members of Wellness Councils while they round out their own in-house efforts.

Part II outlines how the Wellness Council of the Midlands started and gives insight into how you can start one in your own community. This approach to worksite wellness has become a model that is working in cities across the United States. Many cities are forming councils as this book goes to press.

Part II also takes you through the day-to-day operations of the successful nonprofit corporation—the Wellness Council. From this blueprint, any business leader can tailor a similar Wellness Council in any city in the United States.

Wherever you begin, if you and your company are involved in worksite wellness, you are on the cutting edge of an exciting movement destined to leave its mark on American society for years to come.

Wellness in the year 2000 is discussed in Part III. The insurance industry, which has taken a leading role in promoting worksite wellness, is planning to take even bigger steps toward creating a smoke-free society, forming a network of Wellness Councils, and rewarding people who are healthy by charging them less for health insurance.

Some observers thought jogging was a passing fancy. It's not. All those sales of athletic shoes and sweatsuits turned out to be a major part of the revolution in health and fitness in the last half of the twentieth century. Even pollster George Gallup observed that no other change in American behavior had been so profound. People are living longer than they did at the turn of the last century, and by the year 2000, Americans will have added more than four years to their life expectancy.

Although the effects of AIDS or other as-yet-unknown maladies on the length and quality of life are unknown, we Americans are entering the Age of Lifestyle. How we live will determine more and more how well and how long we live. Medical science will tame the diseases we used to fear, but only we can take personal responsibility for our own health and well-being. At the same time, however, rising health care costs have triggered massive changes in the way our health care is delivered to us.

In 1984 alone, more Americans lost their lives to cigarettes than died in all of World War II. Isn't it encouraging to know that when historians look back on the decade of the 1980s, they'll respect another kind of war—one whose mission was to enhance humanity's well-being, not to destroy it.

This new kind of war would be fought by volunteers of all races, by men and women, rich and poor, young and old. Collectively, they would grow into the largest army the world had ever known. The enemy would be the killer diseases of the 1980s. The only weapon would be lifestyle, and the victory would be measured one person at a time as millions upon millions of individuals began to take personal responsibility for their own well-being.

The battle would be fought on jogging trails and in swimming pools, in nutrition programs and smoking cessation clinics, in stress management seminars and in Weight Watchers and Alcoholics Anonymous. Victory would eliminate the number one cause of premature sickness and death because the main killers—heart disease, cancer, stroke, and accidents—are diseases of lifestyle.

Health promotion, behavior modification, lifestyle change, smoking cessation, personal responsibility, healthy lifestyles—these are today's buzz words. If they become tomorrow's arsenal, we will eliminate our greatest enemy. We will win our greatest victory.

William M. Kizer

Omaha, Nebraska
September 1986

ACKNOWLEDGMENTS

My three co-founders of the Wellness Council of the Midlands (WELCOM), Robert Daugherty, chairman of the board and chief executive officer of Valmont Industries; John Kenefick, chairman of the board of Union Pacific Railroad; and V. J. Skutt, chairman of the board of Mutual of Omaha, reacted quickly and positively in support of my vision of corporate wellness. But, even more, they gave of themselves during the early organizational stages, which included making public endorsements, hosting our kickoff luncheon for executives, and providing personal guidance. Without them, WELCOM would not have succeeded in Omaha.

Fred Schott, more than any other single person, bought into the concept and gave selflessly of his energy and talent. Fred served as the first president of WELCOM. An accomplished public speaker, Fred has written and published delightful and helpful books on family life. His many skills as a discussion leader, as well as his ability to conduct seminars and meetings, has been a major factor for WELCOM's success. Fred is a contributing author to this book.

WELCOM's first executive director, Karen Murphy, brought qualities of fun and enthusiasm to WELCOM's activities and was the spark plug of our young organization.

When Karen left to head the health promotion efforts for Southwestern Bell in St. Louis, Harold S. Kahler, Jr., became WELCOM's second executive director. Harold is a totally professional health promotion administrator and brought many new talents to WELCOM as it began its third year. The business plan for health promotion introduced in this book is Harold's special contribution.

The insurance industry has become a major source of horsepower for the creation of community Wellness Councils. Jerry Miller, the former director of public relations for the Health Insurance Association of America (HIAA), has been a prolific writer and strategizer in support of the wellness movement. With Harold Kahler, Jerry was a key player in helping to bring a national conference dealing with corporate wellness to fruition.

The strategy for this landmark conference was orchestrated mainly by Cranston Lawton, former communications vice-president for Aetna. Cran has since retired from Aetna and is now a consultant to the HIAA, helping to form and guide new Wellness Councils. We thank Cran for reading early drafts and keeping our writing on track.

Kenneth Higdon, former president of Business Men's Assurance Co. of Kansas City, took early retirement and became a paid consultant to the HIAA. In this book, Ken tells us how he appeals to other insurance company presidents to become active in the Wellness Council movement.

Stan Karson, executive director of the Center for Corporate Public Involvement, a subsidiary of the health and life insurance industry, provided much assistance to Omaha's first effort, having convinced the White House of the merit of this community initiative and securing a letter of endorsement from President Reagan. No less a believer in the merits of Wellness Councils has been Jim Moorefield, president of the HIAA, who has repeatedly brought recommendations to his board of insurance company presidents and chairmen urging continued and expanded allocation of resources in support of wellness endeavors.

James Brennan, vice-president for Northwestern National Life, and currently serving as chairman of the HIAA's health education committee, has been a prime force for wellness at his own company and in his community, Minneapolis. Thanks also to John Creedon, president of Metropolitan Life, for continuing his company's pioneering efforts in wellness.

Omaha went public with its goal of becoming the wellness capital of the world in January 1982. The out-of-town expert who really sold the idea to the city's business leaders was Charles Berry, M.D., former chief medical director for the NASA space program. In promoting corporate wellness, Chuck drew on his experience of getting man to the moon and back to demonstrate the logic. This book recounts the salient points Chuck cited during Omaha's kickoff and the points he's still making as new Wellness Councils get started.

The most helpful person to me in convincing V. J. Skutt, chairman of Mutual of Omaha, to sponsor WELCOM was the former Surgeon General of the United States Air Force, Gen. Kenneth Pletcher. General Pletcher, a doctor and advocate of health promotion, validated my claims concerning the merits of a community-wide corporate wellness initiative.

Other physicians who sided strongly with me before corporate acceptance was established include Robert Murphy, a pediatrician; Jim Kelsey, medical director for Northwestern Bell; the present and former vice-presidents for the school of health sciences at Creighton University, Richard O'Brien and Robert Heaney; and Alastair Connell, former dean of the University of Nebraska Medical School and now vice-president of health sciences at the University of Virginia; Robert Eliot, author of a best-selling book on stress and director of preventative and rehabilitative cardiology at St. Luke's Hospital in Phoenix; and Calvin Fuhrmann, chief of the respiratory division at South Baltimore General Hospital. Health educators Dick Flynn and Kris Berg of the University of Nebraska's department of health, physical education, and recreation have provided technical advice to WELCOM before and since its inception.

Christina Montgomery of the Health Planning Council of the Midlands was the first person to broach to me the idea of forming a corporate venture to promote healthy lifestyles. She must have been very persuasive since my first reaction was one of skepticism.

It goes without saying that the original members of WELCOM's board have contributed in many ways to this book. Those people include Bob French and Elizabeth Burchard, who were members of WELCOM's original formation committee and even handled such nitty-gritty chores as writing articles and bylaws for WELCOM. Greg Jahn has served continuously and provided sound guidance to WELCOM since its incorporation, in his role as corporate secretary and attorney. Lou Bradley produced 12 public service announcements—which garnered $237,000 of free air time during the first year they ran in Omaha. Frank McMullen, formerly an admiral in the United States Navy and now Omaha's full-time Chamber of Commerce president, invited WELCOM to set up office in the Chamber's handsome new quarters. Tom Whalen of Valmont Industries helped lead his company to a highly successful level of wellness activities, making Valmont a national model. Treasurer Kathy McCoy kept WELCOM solvent, and nurse Beth Furlong of Creighton University headed WELCOM's original liaison committee. Ed Pugsley headed WELCOM's employer assistance committee. Bob Murphy and Jim Kelsey have each headed WELCOM's medical advisory committee, ensuring that all WELCOM's activities were medically sound.

Michael Morrison, S.J., president of Creighton University, has supported WELCOM since its inception and has encouraged a variety of health promotion activities on Creighton's campus at the medical school and in various undergraduate curricula, not to mention wellness programs emerging for both faculty and students.

Throughout WELCOM's infancy, one man, an Omaha physician and marathoner, Eugene "Speedy" Zweiback, has provided the credibility and energy a new organization like WELCOM needed. Today Speedy is WELCOM's chairman and spokesman and, in my judgment, Omaha's best example of a really "well" individual.

In the true spirit of coming together for a common goal, the WELCOM member companies supplied information about their programs for this book. With their overwhelming support and encouragement, I am able to share many good ideas for programs any company can do. The WELCOM members represent a cross-section of American business and industry, blue-collar and white-collar workers, men, women, clerical, union, and laborers, in other words, all sectors of the American work force.

Michael Fortune, a senior partner in the firm of Erickson and Sederstrom in Nebraska, provided the information dealing with the legal aspects of corporate wellness found in this book. Len Pacer helped us interpret the tax implications of wellness programs for corporations. Formerly with Touche Ross, Len joined my company as a vice-president. Thanks to them, the legal and tax sections of this book are the most complete and astute discussions of these topics to date.

The recommended reading section was assembled by librarians Karen Hackleman and Kelly Jennings. Karen was formerly librarian for the medical school at the University of Nebraska and is now located at the University of Maryland Health Sciences Library, where she is the consultation coordinator for the Southeastern/Atlantic Regional Medical Library Services. Kelly is the health information librarian for the Tulsa City-County Library System in Oklahoma. Special thanks go to Father Flanagan's Boys' Home for donating library services for the research of this book. Donna Richardson, head of Boys Town's search service, deserves recognition for her ability to find what we needed in the computer data bases.

Four health professionals and business colleagues served as critical readers, and for their expertise in helping us shape a useful tool for business leaders, I thank Richard Bellingham, president of Possibilities, Inc., a consulting firm in Basking Ridge, N.J.; Roger Heitbrink, a market researcher; William Baun, manager, health and fitness, Tenneco, Inc.; and Lynn F. Bardele, manager, personnel relations, Northwestern Mutual Life.

If Charlie had his angels, I certainly had mine—most notable being my secretary, Audrey Owens, who scheduled meetings, served breakfasts and lunches, and stocked the beverage bar for late afternoon meetings, not to mention acting as hostess at meetings, supervising mailings, managing my schedule, and even managing me when things got hectic. The other angels to whom I am indebted included Jean Hempel, Cleo Ellinger, Betsy Murphy, Kristin Edwards, Sally Lorenzen, Kathy Castilow, Lisa Headley, Sue Brookhouser, and my own daughter Lucy Kizer Smith. Within my company, I also want to single out a wise man, John Mace. He may have retired from the insurance business, but he quietly and competently uses his sales skills to bring new companies into the WELCOM ranks. All of these angels have been consultants and advocates of WELCOM.

Every author should have an editor who is as enthusiastic and competent as Janet Walsh Foltin at John Wiley & Sons.

As part of my writing team, along with my contributing authors, Fred Schott and Harold Kahler, came Sandra Wendel, who put the whole project together. Sandy convinced us that we had something to say and took us through the publication process from proposal through draft after draft to bound book. Sandy's contribution to the writing and editing of this book is evident on every page and in every word.

Without all those I've mentioned by name, and an army of delegates and believers that I don't have room to acknowledge, WELCOM would not have happened and this book would not have been written.

I would like to express my affection and gratitude to them all and wish them only wellness.

W. M. K.

CONTENTS

THE HEALTHY WORKPLACE

INTRODUCTION: THE VISION

American industry can't afford not to expand the wellness movement in the workplace. . . . We need to go with the prevention over the cure. We need to get down to our fighting weight and explore every opportunity we can to hold our own against the competition— especially the foreign competition. The bottom line is, we can only be as good as our people. So if we're to keep our competitive edge in America, our employees of all ages have to be healthy.

—Roger B. Smith, Chairman of General Motors
Remarks at Conference on Worksite Health
Promotion and Human Resources:
A Hard Look at the Data, October 1983

THE GENESIS: "WHY ARE PEOPLE DYING PREMATURELY?"

My interest in wellness originated almost 20 years ago when I was concerned about our claims and our loss ratio. My father, T. L. Kizer, had started Central States Health & Life Co. in Omaha in 1932, and I had worked myself into a position of responsibility.

I sat down with some claims adjusters, and as we pored over the claims, I asked, "Why did they die prematurely?" "Why are these people in the hospital?" And the answer was that most of them were there because of the life that they led. These things were self-induced. They drank too much or smoked or were overweight.

Now that was not in any way a scientific study, but research has since proven those observations to be correct. If we've learned anything about illness and wellness, it's that more than half of the people in this country go to a doctor or enter a hospital because of their life-style and not because of something that was genetically induced. That was the genesis for me.

So what started out as perhaps a self-serving motivation of an insurance executive ends up being something that's not only good for business but, even more important, good for humanity as well. It is in the interest of all society and all humanity to improve the quality of life and to prolong life. One of the by-products is that people end up paying less for their insurance and living longer.

We in the insurance business have a selfish interest, of course, in keeping our clients healthy. If you're healthy and continue to pay your life and health insurance premiums, we make money. If you have medical expenses or die, we pay you, your doctor, the hospital, and your beneficiaries.

At the same time, though, if health promotion keeps people healthier (and it does), then employers will pay lower insurance premiums, and health care costs will stop rising excessively. Everyone benefits—especially employees.

Yet the tangible results of worksite health promotion are hard to measure. You can count the days employees are absent and compare this figure with previous years. You can compare the cost of insurance premiums from year to year. But insurance mortality tables offer even stronger statistical evidence that we are making progress and extending our lives.

An Actuary's Pipedream

Let's look at the actuarial table for 59-year-old men—a group I'm particularly interested in. In 1974, 1,000 of every 100,000 men in that age group did not live to celebrate their 60th birthday. Today, our actuaries figure that only 700 out of 100,000 will not live to age 60. That's a gain—and a significant one—especially if that 59-year-old man is your husband who smokes or a key executive in your company who likes to overeat. It might even be you.

Something is keeping these men alive longer. Certainly some are exercising now; others are eating an average of 10 to 15 percent less fat and cholesterol; half of them are controlling high blood pressure through lifestyle changes or medication; and many of their employers are insisting that they enroll in stress management programs in which they are taught to relax.

If this group of Americans can make such a dramatic impact on actuarial tables in a decade, imagine what significance healthy lifestyles will have for all men and women at every age. It's an actuary's pipedream. Because we've added four years of longevity to the mortality tables since 1970, insurance companies will collect premiums that much longer before people actually die. By the same token, insurance companies are already factoring the savings into the premiums people pay.

Business leaders say to me, "Kizer, you're just into this wellness movement because your insurance company will make money." My response is that that's what American business is all about. It should be sufficient to appreciate the gift of a longer life—a gift from yourself to yourself or from employer to employee. It's good for business and it's good for humanity.

Your Employees Want to Be Healthy

Although I couldn't do anything 20 years ago about the lifestyles of our company's policyholders, I could do something about my lifestyle and the lifestyles of our employees. It's the old saying, let there be good health, but let it begin with me. I'm not a nut, and I'm not eccentric. I don't practice all of the different things I could, but when it comes to the big things, I'm careful. I used to smoke, but I realized it was bad, so I quit. I learned that stress is something you can control. You've got to listen to your body.

In 1972, millions of Americans read about a study conducted by researchers Belloc and Breslow. They outlined seven simple things a person could do to maintain good health:

1. Eat three regular meals each day with no eating between meals.
2. Always eat breakfast.
3. Exercise moderately two or three times a week.
4. Sleep seven to eight hours at night.
5. Do not smoke.
6. Maintain normal weight.
7. Drink alcohol only in moderation.

I was personally interested in my own well-being and fitness, and I was president of the company. I thought, if it's good for me, maybe it will be good for other people too. Even back in 1959, when we moved into a new Home Office, we designed a modest fitness room for men.

By jogging or riding an exercise bicycle at lunch or right after work instead of going to a bar and having a couple of drinks, you're getting a better alternative to the "happy hour."

Bear in mind that the fitness room was for a limited number of executives—all men—at that time. And there was no opposition to it. At that time it was considered to be an executive perk. It was only open to roughly a dozen people, but more than half of them really used it. The biggest surprise to me as I look back was that it was the best grapevine we had. We conducted a lot of business in that room. It was a great communications tool.

We're such busy people, I wondered if it would be possible for people to accomplish wellness activities outside of work. I realized that, for some, the only time to do something was during the workday, and the only place was at work. I strived to create a climate that supported healthful activities.

Women especially are so busy juggling work and family. They believe in exercise and good diet, but they have to struggle to fit it into their lives. Because of the healthy climate within the company, it wasn't long before some of those time-pressured women formed a jogging club and mapped out a course that took them a mile or two away from the Home Office building and back. In the winter they were doing warm-ups in the conference room. The men had their own fitness room and showers, but the women had not yet asked for equal facilities. It dawned on me that here was something I could provide. Today the women have their own facilities—we rearranged some offices for that—and they have invited the men to join them in aerobic workouts. In smart companies, wellness activities are not just executive perks anymore.

Today when I speak to industry leaders, I make it simple to begin wellness for any CEO. I call it the SANE approach. An expanded discussion of this appears in the first part of this book. Number one, I tell them, issue a policy on smoking. Anybody can sit down and write a policy. It takes a little guts, but do it.

Smoking was costing us at Central States in a lot of measurable ways. But also consider what it's costing the nonsmoker who must work in a smoky environment. And the added maintenance on filter systems and air-handling units and upkeep on cleaning glass—you can't put those costs into dollars.

I drew the lines in my own company, but I also knew that at least 70 percent of the people didn't smoke, so I immediately had a plurality. But the biggest surprise to me was that I was a popular guy after I issued that policy. No one left the company because he or she couldn't smoke at the work station.

Instituting a smoking policy is an easy choice for employers, says John Pinney, executive director of Harvard's Institute for the Study of Smoking Behavior and Policy:

> They can do nothing and wait to be sued by an employee with a rightful claim for injury from exposure to tobacco smoke. They can wait until the local, state or federal government requires them to act. They can continue to pay for smoking in a variety of direct and indirect ways. Or they can take the easy, low-cost option of adopting a policy on their own. The

evidence seems clear that the latter option is not only the sensible one but also the right one.

The second simple and no-cost way to start wellness at work is to write a policy on the use of alcohol at company-sponsored events. Like smoking, alcohol abuse drains a company's resources—both human and financial. At my company we rarely serve alcohol at business luncheons, and at company dinners we give guests the option not to drink at all by having soft drinks and fruit juices prominently displayed. It's part of creating a healthful corporate climate.

Number three: If you have a food catering service, have them post the calories on each of the items on the menu. Or ask the vending service to stock the machines in the break room with more healthful items like nuts, sugarless gum, popcorn, and fruit—and leave out the chewy gooey chocolate bars.

The fourth area is exercise. By allowing aerobic workouts to take place, to clear the conference room at noon or after work for stretching, you again add to the healthy corporate climate—and the rest follows.

If you do nothing more, you've got a good wellness program, and you can start right away. You'll find, however, that employees are hungry for more, and they will move on to other activities. Things get organized, and record-keeping begins. People feel good about themselves, and the impact on morale becomes obvious to everyone. Employees take pride and ownership in their activities. Management's role is simply one of providing support and drawing reasonable limits.

Even if you've joined the ranks of true believers that worksite wellness is not here today and gone tomorrow, keep reading. The next chapter answers the tough questions for business and industry—the same questions we kept asking ourselves while we were wondering if we were chasing rainbows.

"Only as Good as Our People"

Instituting wellness programs in your company is part of a three-pronged approach to getting health care costs under control. Promoting health at work to lower the need for health care is the key strategy. The other two areas are (1) revamping health insurance policies and shifting costs, for example, by increasing deductibles, or having employees pay more for health insurance coverage; and (2) changing the structure of the health care delivery system by introducing those relatively new, and not-yet-fully-understood organizations called health maintenance organizations (HMOs) and preferred provider organizations (PPOs).

By using one of the above methods or some combination thereof, a company performs major surgery on its health care picture. Reducing deductibles and shifting costs to employees works immediately. But the only long-term solution is the thinking person's answer to cost containment, and possibly even cost reduction, and that's worksite wellness.

It's hard to go out and find good companies that aren't doing something about wellness right now. I can't envision a board of directors 10 years in the future wanting to hire a chief executive who wasn't a good wellness role model, because the success of the business is now at stake. The worksite wellness movement is based on individual choice and responsibility, which is the heart and soul of American culture. Because we're competing in world markets, President Reagan has said, "America can only be as strong as its people."

If worksite wellness continues to grow, and if millions of employees and their families can be helped to help themselves achieve better health and healthier lifestyles, then in the year 2000 and beyond, the 1980s will be seen as the decade when American businesses began to understand the important link between health and work.

Even the most skeptical business leader can be convinced that wellness flows through to the bottom line. I have always said, "There is no red ink in wellness." I also knew that health promotion movements didn't get anywhere unless direction came from the top. That's why I approached it as one chief executive officer to another.

By recruiting three of the most prestigious executive officers who represented the major areas of industry in the Midlands to join me as founders of the Wellness Council in 1982, I knew we could convince other corporate officers to join us. It didn't matter if a company had started a wellness program—many didn't know what wellness was yet—but all of the charter members realized that worksite wellness was an idea whose time had come. Individually, we could do something in our companies. Together we'd create an even stronger movement in our companies and in our community.

THE WELLNESS COUNCIL: A COMMUNITY INITIATIVE

I want to give you a brief history of how the Wellness Council of the Midlands (WELCOM) got off the ground in Omaha. Although this isn't a historical account, a short summary of the events that led up to the formal incorporation of WELCOM in 1982, and the thinking that went into the movement, will be useful for those of you who intend to take the lead in your own communities.

I knew my efforts in my own company were on track after I attended a conference on health and wellness sponsored by the Health Insurance

Association of America in Atlanta in 1980. I remember flying home from the conference with Dr. Robert Long, the medical director of Mutual of Omaha, one of the corporate sponsors of the conference. Dr. Long told me in no uncertain terms that the future of medicine was in prevention and in the individual taking charge of his or her own health. I became a true believer.

In 1981 the Health Planning Council of the Midlands—a health systems agency—was trying desperately to contain health care costs but not succeeding. And their federal funding was running out.

So when a member of that agency heard about the wellness activities at my company and asked if I would help them start a community initiative involving business, I was anxious to do it. It sounded like an impossible job, but I went to a meeting. Also in attendance were six or seven other people from the community, including a professor of health education, executives of the phased-out federal health agency, and personnel workers from the phone company. It was a diverse group.

The most productive thing we did was hammer out a statement of purpose saying that our mission was to foster wellness at the worksite. Our strategy was to gain the active support of the chief executive officers in the business community.

Meanwhile, I was preparing to ask the board of directors of my company, Central States Health & Life, to put a running track around our new Home Office building. I thought the idea had merit; that it was good for morale, good for fitness, and good for the company's image. I'm not a runner myself, but the employees were interested in having a track rather than running on busy streets near the building. And the architecture of the building easily lent itself to a circular track. We'd open it to the residents of the surrounding suburban neighborhood— the early-morning runners, afternoon walkers, and kids on bikes. Certainly there was PR value for the company in that invitation.

Because this was a community initiative involving the insurance industry, the Center for Corporate Public Involvement (an insurance industry group) agreed to pay for a consultant to come to Omaha and present the case to the board. Seizing the opportunity to do even more public relations, I invited some people from the health planning group and from some local companies to hear the speaker who was Dr. Jonathan Fielding, a private consultant to business and a professor at UCLA. After Dr. Fielding's presentation to the board, we had lunch and discussed what was going on in the worksite wellness movement.

When the meeting was over, several people said, "Why don't we get together for a couple more of these?" The feeling was that what was happening at Central States and a few other companies was important. Here were local companies that cared enough—for a number of reasons—to actively promote wellness and share their ideas with other, often competing, companies. Perhaps, they thought, other companies should be doing something too.

That was the cement that held things together. The idea had been planted in that early meeting with the federal agency people, and a simple mission statement and strategy had been drawn up.

My board, by the way, agreed to build the track. But the rest of us met a couple more times and decided to form a council. We outlined some elementary bylaws and articles of incorporation and plunged forward—never certain where this whole thing would lead—to recruit support from the chief executive officers of Omaha's biggest and best companies.

My next task, then, was to contact strong, civic-minded CEOs. I chose three of them who were also my personal friends. I called Robert Daugherty, head of Valmont Industries, the world's largest manufacturer of center-pivot irrigation systems, and a major employer in the Omaha area. After I gave Daugherty a brief description of the proposed Wellness Council, he said, "As far as I'm concerned, you have an open checkbook. Good luck and keep me informed."

Today, Bob Daugherty affirms his long-time commitment to promoting healthy lifestyles for his work force, and his company's health care costs have leveled off and started a downward trend in times of rising health care costs nationwide. As one of the founders, Daugherty continues to credit his company's involvement in the Wellness Council: "We believe that an investment in wellness does indeed pay off at the bottom line. To this end the role of WELCOM in providing leadership and a strong network of support for its membership in the development of wellness programs is a very important service to our business community."

To get further support for the Wellness Council in its early days, I called John Kenefick, CEO of Union Pacific Railroad, and asked for his support. He agreed. Because of the nature of the business, railroaders are of course interested in safety issues, and the health and well-being of the work force complement any efforts to promote safety in the workplace. John Kenefick, as past president of the Omaha Chamber of Commerce, was also interested in the idea of a Wellness Council because, as he says, "a community so visibly concerned with the well-being of its work force has to be an attractive plus for companies and individuals considering to locate in Omaha."

The third CEO was a bit harder to convince. But V. J. Skutt, the chairman of one of the largest insurance companies in the world, Mutual of Omaha, not only agreed, but asked me what else he could do to help the Wellness Council. I asked if he would sponsor a luncheon in his company's executive dining room for chief executive officers only, cohosted with Daugherty and Kenefick. He said yes. He also agreed to send his plane to pick up the guest speaker.

My colleague, V. J. Skutt, brings the ideas of wellness and individual responsibility into perspective when he says, "While we strive to improve the longevity of life, let us remind ourselves that we can improve the quality of life by encouraging wholesome, sensible lifestyles." He goes on, "The Wellness Council provides the vehicle; it is up to the individual to sit in the driver's seat."

The Big Three

The trust of my business colleagues meant a great deal to me. Few names, if any others, carry the distinction in our community as those of Bob Daugherty, John Kenefick, and V. J. Skutt. In Omaha, these three business leaders represent the major sectors of the economy in the Midlands: agribusiness, transportation, and insurance. They encompass manufacturing and service sectors of the economy, union and nonunion laborers, and blue-collar and white-collar workers.

Without their support, both active and financial, WELCOM would not have been formed. Somehow our business sense in that planning meeting months earlier had been right on target: Involve the CEOs—that's where decisions are made.

The Kickoff Luncheon

Omaha's often unpredictable winter weather cooperated that January day in 1982 as 43 CEOs of Omaha's biggest and best corporations gathered for the luncheon at Mutual of Omaha's world headquarters in midtown Omaha. The governor of Nebraska came, and so did a congressman. And Dr. Charles Berry, former medical director to the space program, was the guest speaker. There had never been an event like this before. Security was tight, but amazingly no one was called away to attend to business elsewhere. Everyone stayed for the entire program.

Lunch itself was a lively affair. Though surrounded by the governor, the congressman, and the CEO co-hosts—the leaders who struggled daily with the operations of the state, the country, and megabuck corporations—Dr. Berry was asked not about the risky business of getting humans back from the moon, but about the intricate logistical process of going to the bathroom in weightless space.

Dr. Berry said NASA didn't have all the data to figure out if astronauts could get to the moon and back. In the same sense, he said American business leaders don't have all the data they'd need to decide if they should institute wellness programs. But we'll never have all the data, Dr. Berry pointed out, and business executives are used to making critical decisions without knowing all the facts.

"Healthy, happy, and productive employees are a company's greatest asset on their balance sheet," Dr. Berry told the roomful of chief executives. "Some CEOs recognize this and say, 'Whether it saves money or not, I believe it is the right thing to do for our people.' Others have said, 'We are cutting back so I can't start a new program now.' In fact, you can't afford not to start it, for the benefit of your employees, their morale, and for the future of your business!"

As I look back, I remember thinking that many of us knew each other. Some of us were competitors in the marketplace both locally and nationally. All of us were community and civic leaders as well as business leaders. And smaller groups of us met periodically at Rotary Club and Kiwanis and the like. However, no one could remember another luncheon that was held exclusively for the chief executive officers of the community, and there was an unexpectedly high acceptance of it. Never had we all been together in one room like that before, and we probably never will again.

John Kenefick closed the luncheon with these words: "We think that the idea of a Wellness Council is a good idea. Good health is good business. We hope you agree. You can vote yes by sending us your check and naming a delegate."

By the time the first delegates' meeting was held six months later, even though the only publicity was by word of mouth, 53 local companies had joined—even more than the number of CEOs who had attended the kickoff. More important, those delegates represented over 55,000 employees. Because of the overwhelming support and the apparent success of our kickoff luncheon, the formation committee formally incorporated the Wellness Council of the Midlands.

An Epidemic of Wellness

Central States hosted the first delegates' meeting at our Home Office in July of 1982. I told the 70 delegates in attendance, "The wellness movement is spreading so fast, it could be the most significant epidemic of the 1980s and 1990s." I was appointed the first president of the group; Fred Schott, an independent consultant whom I later persuaded to direct training and development for my company, became the Wellness Council's vice-president.

I am often asked, "You're in the insurance business. You have a company to run. Why are you doing this?" I answer that what started out as something that didn't take much time ended up being a bit of a monster. Because I was running a company, however, I was able to contribute resources, personnel time for mailings, secretarial time for preparing letters, and typesetting and printing for publications.

Of course, I had doubts all along the way, but one thing led to another: The failure of the federal health agency led to a community takeover of its duties; the informal meeting of the health planning council in a back office of a shopping center generated interest in the community; my intention to install a running track led to having a guest speaker on wellness at a board meeting; this in turn led to the formation of a committee, to a formal statement of purpose, and to a luncheon hosted by Omaha's biggest and best corporate executives. The rest is not only history but the birth of the first Wellness Council.

WELCOM proudly claims to have over 112 corporate members representing 70,000 or more employees in the Midlands. The number of delegates grew from the original 70 to over 210, and more companies join every month.

The insurance industry, and specifically our trade associations, the Health Insurance Association of America (HIAA) and the Center for Corporate Public Involvement (an entity of the HIAA and the American Council of Life Insurance), are actively pursuing worksite wellness. They also put their money where their mouths are by getting similar Wellness Councils in other communities off the ground. It is appropriate for the insurance industry to play an active role. Because of their efforts, Wellness Councils are now springing up throughout the United States.

To show how far the Wellness Councils have come: In October 1985 the Department of Health and Human Services, in joint efforts with the Health Insurance Association of America, launched a program to support a network of employer-sponsored Wellness Councils in cities across the country. I was proud to attend the press conference in Washington, D.C., and to stand alongside 10 other business leaders representing the 10 other Wellness Councils in existence or being formed at that time. WELCOM's current chairman, Dr. Eugene "Speedy" Zweiback, accepted the charter founder certificate—the Omaha council was singled out as being the model for the others and for future councils.

The vision had come full circle.

PART ONE

1

FITNESS THRIVES FROM 8 TO 5

As a nation of farmers-working in the fields all day and, later, as a nation of laborers also using our bodies to work, we had little need for or interest in going to a fitness boutique at the end of the workday. But now we are a nation of clerks; we use our heads all day, not our bodies. In that context, it is easy to see that fitness is not a fad and to see why it is appropriate that corporations become deeply involved in health and fitness.

—John Naisbitt and Patricia Aburdene
Re-inventing the Corporation

The concern for personal health and well-being did not, of course, begin in the workplace. What you are seeing in your community parks and along the roads is more than a jogging craze. Sporting goods stores are selling a lot of running shoes, weight machines, warm-up suits, and books on fitness and health. Across the nation, health food stSres and health clubs are springing up. Workshops, seminars, and classes on everything from stress management to biofeedback, child rearing, nutrition, and smoking cessation are being sponsored by all kinds of groups. Even a number of for-profit entrepreneurs are joining the effort to help people get and stay well.

John Naisbitt, in *Megatrends* (1982), and other observers feel that the growing emphasis on health and fitness was born in the counterculture of the early 1970s. Getting "in touch with yourself," "returning to nature," and "finding your roots" were hallmarks of the "me" generation—not the kind of hard-headed stuff that appeals to business leaders. But the sometimes narcissistic search of the 1970s has helped to feed the healthy movement identified by Naisbitt as the trend toward self-help and accepting personal responsibility for one's own well-being.

There is a growing body of evidence that the new-found interest in health is paying off in more than just dollars. One area of statistical evidence is the insurance industry's mortality tables. People are simply living longer—four years longer according to recent figures—and that's the biggest gain since medical science tamed polio and tuberculosis.

Reasons for such a dramatic gain go beyond the fact that a great many more middle-aged executives are exercising. We also know that people are, on average, eating 10 to 15 percent less fat and cholesterol. Fifty percent more of them are controlling high blood pressure through medication. Many of their employers are insisting that they enroll in stress management programs where they are taught to relax.

Although somewhat belatedly, American business and industry have gotten into the act. They are not just dabbling in it either. Worksite wellness has already taken root and will be a part of the workplace for decades to come. Already hundreds of companies—at least half of the Fortune 500—are putting some of their profits back into their people. These activities range from a pickup game of volleyball in the parking lot at lunch to sophisticated on-site facilities with well-trained staffs who carefully design comprehensive programs.

The worksite wellness movement has taken off so fast that many business leaders, left behind in the puffery of a smoking policy, are asking: "Why the workplace?" and "Why now?"

Many CEOs and company presidents have personally discovered the value of fitness and health—the hard way. Many learned to cope with the rat race and pressures of modern business through regular routines of physical exercise, watching their diets, learning to deal with stress, and paying closer attention to their family and other important nonwork activities—*after* their heart attacks.

Regrettably, much of the impetus for these executives came from near-tragic events resulting from poor lifestyle choices. They learned to lose weight and found they had more energy, felt better, and even experienced a rededication to their careers. These executives concluded that if it worked for them, it should work for their employees. They turned into apostles of the benefits of personal health and well-being.

Seldom does a day go by when chief executive officers are not solicited with phone calls, letters, and special appeals, all asking for money for a worthy cause. But worksite wellness is not a charitable cause. Supporting the movement, in your company or within a national industry, should not be seen in the same light as charity. It may look similar on the surface, but supporting wellness is not the same as joining the arts council, making a contribution to help people caught in poverty, or giving to any other good cause.

In the same light, worksite wellness is not just another fringe benefit for employees. It may, unfortunately, become a part of the negotiation dialogue between labor and management, in which both management and labor might view worksite wellness as a bargaining chip that can be put on the table today and given up tomorrow for another advantage.

Labor has a right to be suspicious of wellness when it is used as a bone thrown to dogs in exchange for so-called real health benefits. Management has a right to be suspicious of frivolous requests that amount to country clubs for working men and women. Worksite wellness should not be seen as a frivolous extra or the gift of a generous CEO.

But General Motors, with Roger Smith at the helm, hammered out a contract with the United Auto Workers union that marked the turning point for wellness. GM workers opted for job security over pay increases in that 1984 contract, which also calls for GM to provide wellness benefits. Observers say those wellness benefits (even though they are slowly evolving into a full-blown program) are a milestone in labor and management relations. The monumental agreement between GM management and labor brought wellness out of the realm of "executive perks" and put it into blue-collar terms that business can understand.

TOUGH QUESTIONS FOR BUSINESS AND INDUSTRY

Meanwhile, across the street from GM, Chrysler was trying to get a handle on rising health care costs. When Chrysler chairman Lee Iacocca asked Joseph A. Califano, Jr., to head up a committee to look into health care costs at Chrysler, Califano eagerly agreed. Califano had served as Secretary of Health, Education and Welfare from 1977 to 1979 and had helped devise some of the Great Society programs including Medicare and Medicaid.

Chrysler was being strangled by rising health care costs—to the point that overuse and abuse was adding more than $600 to the price of each car Chrysler sold. Consumers, who are used to scanning the car stickers for options like AM/FM radio and whitewall tires, would be alarmed to see something like "Employee Health Care Costs: $600" tacked onto the window price. Califano and the Chrysler committee decided to do something about it.

Califano told the Health Insurance Association of America (1985) that "the key to health care cost containment rests in an aroused private sector. . . . The problem didn't crop up overnight. Since World War

II, government, unions and corporate management have focused on providing more and more health care benefits for our people, without realizing they were becoming hostages to costs beyond their control."

Because he was not willing to allow mounting health care costs to slow Chrysler's rise from the ashes, Califano told these insurance executives that the company, like many other businesses, had relied on its insurance carriers for information. "We didn't know how much we were paying, for what medical tests, diagnoses or procedures, to what doctors, hospitals or laboratories, for which employees, retirees and dependents, or how often. So the first step was to get the facts—the kind of information any businessman or individual buyer should get from any seller."

The facts spoke for themselves. Waste and overpayment were rampant. So in addition to making massive changes in health care insurance coverage, Chrysler mounted a major health promotion and disease prevention effort.

What's the payoff for Chrysler? In 1984 their health care costs were reduced by $58 million. As Califano concludes, "Chrysler and corporate America are at last moving and shaking the health industry."

The High Cost of Doing Business Gets Even Higher

Over the past 30 years, health care has continued to be the fastest rising cost of doing business in America. Health care costs are making a bigger dent in an employee's take-home pay than income taxes. On what is our federal government spending more money than defense? Health care costs.

We've been able to put the brakes on inflation, yet health care costs continue to soar out of sight. The cost of medical care continues to double the increase in the overall consumer price index. In 1985 the total bill for health care in the United States was more than $1 billion a day. U.S. companies paid over $87 billion in health insurance premiums for their employees, retirees, and their dependents. That's more than those companies paid out in dividends.

Health care is, of course, big business. Counting all the doctors, nurses, paraprofessionals, and other health care providers, health care is our nation's second largest employer (education is first). And health care ranks third among the things on which people spend the most money (food and housing are one and two).

We are all rightfully concerned about annual budget deficits of about $200 billion. But this nation's annual health care bill alone is more than double the deficit. Just think, if we could cut our health care bill in half, we would save this nation the equivalent of that troubling annual deficit. Impossible? On any given day in America, half the people in the hospital (over age 45) are there because of poor life choices, according to health insurance figures. In other words, they are there because of disease, sickness, and accidents brought on by what they smoke, eat, and drink; how they exercise (or that they don't); how they handle stress, abuse chemicals, or drive while under the influence (and almost never buckle up).

Again, we are beginning to make progress, and not only in the way the government administers Medicaid and Medicare. The preferred provider organizations (PPOs) will, over the long haul, help to reduce our health care bill. The trend in benefit redesign will also help. Plans which trade first-dollar coverage through various deductibles for greater overall coverage for the patient make sense. This is especially true when the plans allow for second opinions and for preventive medicine, like regular checkups.

All this cost shifting will help to slow the spiraling rise in health care costs to American business. But, by themselves, they are not enough. Nothing will slow and even reduce our nation's health care bill like helping more and more Americans take personal responsibility for their own well-being. The key to lowering costs is to keep people from getting sick to begin with.

Worksite: The Logical Point of Impact

More people spend more time at work than at any other activity except sleeping. This fact alone makes the worksite the logical place to reach people, no matter what the message. Work, too, contributes to people's health and well-being or lack of it. Studies often reinforce the idea that "happy workers are productive workers."

Work itself is sometimes its own worst enemy. Pressure for production produces stress. Deadlines and quotas create high pressure, rush-rush environments. Competition with other workers and with other companies also brings with it the stress that kills people. Add to that the general rat race our society is so often accused of being, the increase in working women with young children, and the higher incidence of two-paycheck families, and you have the ingredients for over-stressed, overworked, sometimes underpaid, unhappy, unhealthy workers. And the employer—as well as the worker—pays the price.

Health promotion at work can be considered a "pay me now or pay me later" proposition. The employer either pays for programs that enlighten workers about their own health and what they can do about it, sometimes at the employers' expense, or the employer pays for medical diagnosis and treatment after the employees get sick. These latter payments are far steeper than the prevention programs. Consider, too, the money employers may spend for workers' compensation, early retirement, and retraining to replace employees who are no longer able to work because of lifestyle-related diseases.

But why would employees be receptive to health promotion at work? They don't always respond to appeals to participate in other voluntary programs such as United Way campaigns or planning the company picnic.

Work, believe it or not, has proven to be the ideal place to reach people. In some programs voluntary participation can run as high as 80 percent. More people take part in health promotion activities at work than they do in community activities such as blood pressure screenings in shopping centers or health fairs. Work is convenient. And a corporate culture that supports participation—and makes time for employees to participate—reaps even higher benefits on the bottom line.

Carry the idea of support even further. Employees who smoke together in the designated smoking areas and who are given the chance to stop smoking at company expense may support each other in their individual efforts. The same is true with weight loss programs. Misery loves company, and workers will take advantage of the social support network if the people they work with every day are going through the same behavioral changes.

Worksite wellness programs give employers a chance to keep the work force in tiptop shape because doing business in America—like the stock market—has its ups and downs. A work force that can physically and mentally withstand whatever the business climate brings can weather any boom, recession, depression, merger, takeover, or expansion.

Ups and Downs of Doing Business in America

It seems that just when business is great, the market falls off, the product cycle ends, and then there's unemployment. Should we be preparing our work force to withstand the ups and downs of doing business in corporate America?

If employers understand how stressful these situations of uncertainty can be for employees, that uncertainty can create the opportunity for wellness programs to do what they do best—for example, keeping the work force in tiptop shape, helping people monitor their blood pressure levels, giving them opportunities to exercise and stay in shape for the day they can return to work from a layoff, and restoring their self-esteem by letting them feel good about themselves, no matter what happens to their jobs.

Any good, well-managed business will hedge against risks. Worksite wellness is a logical way to hedge against the risk of the work force falling out of good health and into costly and unproductive poor health. Many corporations offer their high-level staff members outplacement counseling services when the company decides these people just don't fit into a reorganization plan. So, too, wellness programs give companies an option when business is in a downturn.

And when business is falling off, companies look for places to cut. Wellness programs might logically be vulnerable to such cuts, but surprisingly they are not. In fact, wellness programs are even more valuable to companies when the storm clouds gather.

AT & T, for example, discovered that the wellness programs instituted during boom times came on especially strong during the divestiture of Ma Bell because workers were concerned about where their jobs were going and what was happening to the company.

Valmont Industries near Omaha keeps putting more and more money into its wellness programs even though the farm situation and downturn in agricultural prices are forcing the company to lay off more and more workers at all levels. Just because farmers can't afford to buy what Valmont makes—irrigation equipment—does not mean that Valmont has forgotten that these times are stressful for the 25-year factory worker who isn't sure he'll have a job next week.

Sometimes companies do have to cut, and wellness programs can fall victim to the budgetary hatchet. But if a company has created a corporate culture that is program oriented, and one in which the employees have a sense of ownership, the company may still be able to provide wellness because wellness need not be a high-ticket item. However, if the program is facilities oriented—for example, the company has hired a wellness director and made an investment in exercise bicycles and weights—then, yes indeed, those personnel are especially vulnerable in troubled times. You don't need five staff members to monitor a pool, check out volleyballs and towels, and clean locker rooms.

A wellness program that creates a healthy environment through smoking and alcohol policies, calorie counts, and blood pressure screenings, and relies on voluntary help from employees to lead aerobics classes or teach CPR, can ride the roller coaster of American business and can even be the strongest department during lean times.

An American Response to the Japanese Challenge

Many books, editorials, and journal articles have been written about the Japanese style of management. Perhaps the most well-known was written by business consultant William Ouchi. It is *Theory Z: How American Business Can Meet the Japanese Challenge* (1981). Although it is just one of several books about the Japanese style of management, Theory Z touched a nerve within American business. A takeoff on the classic Theory X and Theory Y style managers, Theory Z joined them at least for a while as the buzzword for the progressive style of managers in the early 1980s. Many predicted that Ouchi's new theory of management would change the way managers and employees thought about their jobs, their companies, and their working lives.

There is no doubt that in recent years the Japanese performance in the marketplace has caused us to take notice. Ouchi describes in detail the Japanese emphasis on participative decision making, shared corporate values, and holistic concern for people and families. These ideas have been carefully studied and copied by U.S. business leaders.

Even though Ouchi warned against comparing, thoughtful critics of the book began rightfully to maintain that the experiences of Japanese and American business cannot be compared. The two cultures are simply too different. For one thing, the Japanese are a homogeneous society with millions of people crammed into a tiny island nation. The United States is a heterogeneous society spread out over a huge land mass with vastly different territories. The American character is built too much on the ideal of the rugged individual to adopt the Japanese model.

Although Japanese styles of management cannot be simply imported into America like Toyotas and TVs, American business has much to learn from them. The most important lesson is that of holistic concern for workers, which recognizes not only that people have lives outside of work, but also that families are important to everyone. The opposite of this concern is common in U.S. companies, whose managers are often accused of thinking of workers as employees only. The Japanese also recognize that people are social beings who have need of shared decision making and collective responsibility, not only for the job, but for each other. Holistic concern for the individual implies concern for mental, social, familial, spiritual, intellectual, and physical well-being, as well as for the important occupational function. Even when not defined within their cultural context, people are any corporation's most important asset.

Other best-selling books have developed the theme of the importance of people. *The One Minute Manager, Further Up the Organization*, and *The Leader* are just a few. The best seller *In Search of Excellence* and its sequel *A Passion for Excellence* both demonstrate that a concern for people, both employees and customers, has always been a driving force behind successful American companies.

Worksite wellness is a uniquely American response to the Japanese challenge and a natural extension of American respect for the individual.

In Japan, workers generally remain at their work stations or desks and exercise at the same time. In the United States, exercise is voluntary. If a factory worker or vice-president buys into the idea of worksite wellness, that person will find time to fit fitness into a day's work. The fact that the worker and the VP may choose to sweat together, side by side, each day, somehow seems to illustrate the true equality in America. That's part of the theme, developed in this book, that wellness is an individual's responsibility. And wellness is in everybody's best interest.

Wellness Is an Individual's Responsibility

"We allowed our citizens to think that it was somebody else's responsibility for good health—not theirs. It's an individual's responsibility for good health, not the doctor's, government's, or the company's," says Robert N. Beck, vice-president of corporate human resources for Bank of America.

The workplace offers the best place to take on that responsibility, and the employee, working in tandem with the employer, can do things that benefit them both. Employers, by creating a healthy company and an atmosphere that supports healthful lifestyle changes, can encourage employees to take part. Employees, by taking advantage of the atmosphere created by an enlightened employer, can be healthier, and will probably be more productive, absent less often, and have a greater sense of worth to self, family, and company. Both employee and employer will save money on health insurance costs.

This well new workplace is talked about extensively in the popular business books. And in the reinventing of and the passion and the search for the best companies to work for, what is the humanistic theme that underlies the best companies? A concern for people and the creation of a workplace that lets them be at their best. Wellness and health promotion are part of that healthy corporation.

Benefits for Small Business

The Small Business Administration claims that 99.5 percent of all businesses in this country are small businesses. In fact, according to the census, most Americans work in companies that employ nine or fewer people, and half of all employed people work for firms employing fewer than 100 employees. The trend in our service-oriented economy is toward more and more small businesses.

But why have small companies—the real estate office, the boutique, the temporary-services office, the retail clothing outlet—been slow to start health promotion programs in their places of business? For them, margins of profit are measured in thousands of dollars, not millions. So the net effect of an employee or two stopping smoking could mean the difference between profit and loss.

Small businesses, too, are more likely to feel the pinch of rising health care costs, and they are probably out there shopping carefully to get the best buy in health insurance as they look for ways to ease the pain of higher premiums. Health promotion programs are a natural way for them to get an even higher return on dollars invested in their employees' health.

Small businesses are structured less formally, which means there's less red tape between the decision maker and decisions being made. This informality, which often means that employees themselves are involved in decisions that affect them, paves the way for wellness programs to begin once the CEO gives the approval. No one gets bogged down in levels of bureaucracy, running proposals through committees, and delegating responsibilities.

A key strategy for the success of wellness programs is to involve employees in all phases of the decision making. The idea of employee ownership—that the employees have put in their ideas and that the planners themselves are from all levels of staff—sets the stage for employee acceptance of wellness programs and encourages high rates of participation.

A study by HealthWorks (1984) in Seattle revealed that small businesses launch health promotion programs not particularly to save money on their health care costs, but more for human resource management. This reason targets things like morale, absenteeism, and overall health and well-being. Productivity, another component of human resource management, was cited as the second most popular reason that small businesses start health promotion programs. Savings in health care costs were thought to be a by-product of health promotion programs and certainly a long-term benefit.

Probably the most revealing finding from the HealthWorks study was that it doesn't matter how much a company spends for health promotion. In other words, a company that spends $1,000 and carefully plans the programs, targets the problems, and uses the free or low-cost services available in the community can end up with as comprehensive a program as companies that spend thousands of dollars.

The theme presented in this book—that a company can get started in wellness with a small investment or no investment at all—is supported by a detailed business plan presented in chapter 2. Within that business plan are lists of activities that a CEO can do right away to get the company started in wellness.

RETURN ON INVESTMENT

Business can control health care costs and get a return on investment. A company should investigate the benefits of a health promotion program just as rigorously as it would explore the merits of investing money, manufacturing a new product, or changing a marketing plan. But the same CEOs who make risky business decisions every day still want proof that these "wellness things" really work.

Until now that proof has been elusive at best. But the data are coming in, and one company after another is beginning to stand back and say, "We are saving money," or "We reduced medical care costs," or "More than half of our employees quit smoking after we offered smoking cessation programs," or "Employee turnover is down."

Here are some specific statements about returns on investments in wellness:

- In one major insurance company, the employees who participated in a fitness program were only absent 3.5 days; the others were gone 8.6 days. In another similar company, the employees who exercised took only 4.8 sick days a year on average, compared with 6.2 days for those who did not exercise.

- Medical costs dropped 3 percent in 1983 for this manufacturing plant and continue a downward trend because of wellness programs working in concert with a revision of insurance coverage. Turnover rates among all personnel at all levels are also lower (Valmont Industries, Omaha).

- Personalized exercise programs lowered coronary risks for police officers in a pilot test conducted by a university's health science staff (Creighton University program for Omaha Police Department).

- Participating employees collectively lost 285 pounds (Central States of Omaha).

- Over 70 percent of the company's female employees attended a seminar on breast cancer conducted by the American Cancer Society (Metropolitan Utilities District in Omaha).

- A dramatic drop occurred in the number of employees who are late or absent from work. Credit is given to the wellness program (Millard Manufacturing, Omaha).

- This hospital saves $50,000 a year because staff screen job applicants for back injuries and conduct back clinics for employees who are at risk for back injury. Workers' compensation claims dropped sharply (Midlands Community Hospital, Omaha).

- Because the 37-person work force used their health insurance wisely, and because claims were down, the insurance carrier gave the company a rebate on premiums. The company shared the money with the employees (Western Printing, Omaha).

All these claims are documented by the Health Insurance Association of America and the Wellness Council of the Midlands. Now look at the facts about what poor health is costing American business. We have far to go. Table 1.1 summarizes the current numbers.

It's the old business principle: You have to invest money to make

TABLE 1.1. The Facts

- At the turn of the century, pneumonia and influenza were the leading causes of death in the United States. Today, heart disease kills almost 40 percent of Americans, and cancer another 20 percent.
- The only "cure" for heart disease is people taking better care of themselves. The main "cure" for lung cancer is people stopping smoking.
- Some 14,000 person-years are lost each year by private industry in the United States because employees suffer from cancer. That translates into $10 billion in lost earnings and $3 billion in direct costs to business (American Cancer Society, 1981).
- Each employee who smokes costs employers from $336 to $601 (Kristein, 1983) to $4,600 (*Corporate Fitness & Recreation*, 1985) more than nonsmokers yearly. The insurance industry estimates that employers pay $300 more annually in health claims for a smoker than a nonsmoker (*U.S. News & World Report*, 1985).
- The Surgeon General, calling smoking the greatest threat to public health in the country, has challenged the nation to create a smoke-free society by the year 2000.
- Disease and lost productivity due to smoking are costing the economy $65 billion a year—that's $10 million an hour, or $2 for every pack of cigarettes, says a congressional study from the Office of Technology Assessment (*Omaha World-Herald*, September 1985).
- American business paid more than $87 billion for group health insurance in 1984, says the Health Insurance Association of America.
- The largest single supplier to General Motors is not a steel manufacturer but an insurance company. The cost of health care adds $480 or more to the price of a new car, according to GM's chairman Roger Smith (October 1983).
- The nation's medical care bill has grown from $27 billion in 1960 to over $425 billion in 1985. Costs for health care are projected to hit at least $1 trillion by the year 2000.
- Over 80 percent of companies are shifting the increase in the cost of health benefits to employees (*Dun's Business Month*, 1985).
- Americans may be spending millions of dollars on athletic shoes, but a report from the Department of Health and Human Services says that 80 to 90 percent of Americans still do not get enough exercise (*Wall Street Journal*, August 1985).
- Children, too, are less fit than they were a decade ago. Despite the perceived fitness boom, today's boys and girls are fatter and have elevated serum cholesterol levels. Too much TV, and not enough sports and walks to school are responsible (*Omaha World-Herald*, October 1985).
- Over 90 percent of workers will agree to having their blood pressure taken at work, while shopping center efforts net only 20 percent participation.

TABLE 1.1. Continued

- Almost 500 million workdays are lost annually because employees are ill or disabled. Heart disease accounts for 26 million of those days, and back problems add another 93 million lost workdays (C. A. Berry and M. A. Berry, 1984).
- Alcoholic employees and smokers each have twice the absenteeism rates of other employees.

money. So for those who think the idea of worksite wellness sounds good but still aren't sure it saves money (not to mention boosting employee morale, increasing productivity, reducing turnover, and just keeping people well), the facts are in. Business newspapers and executive news magazines are filled with examples from the big companies like New York Telephone, Kennecott Copper, General Motors, Bank of America, Tenneco, Shell Oil, Xerox, IBM, and Johnson & Johnson.

The point is that if these big companies can save money, any company—large or small—can save money. And there's no big business secret to doing so. You make a return on an investment, and it pays off for the company, the employees, their families, your stockholders, and society.

Tax Benefits Versus Perceived Value

Tax law provides some tax breaks for companies that spend money putting in workout facilities. The 1984 Tax Act says that wellness is not a taxable fringe benefit, and that's good news. It's a nontaxable fringe benefit like pension and profit-sharing plans or group life insurance. Whereas if a company pays for an employee to belong to a country club, the employee must in some cases report that as income and pay income tax on it. The cost of that membership, on the other hand, is deductible to the company.

But if a company builds, owns, and operates a fitness facility, under current tax law the company may deduct the cost of operating that facility on its corporate tax return, and the costs are not charged back to the employees as income. A tax adviser, however, cautions that tax laws are subject to change, and a company should get good advice on current issues—although wellness doesn't appear to be the target of proposed changes as of this writing.

Companies might look at a different way to measure the value of a fitness facility or a wellness program. If it costs the company $100,000 to renovate and equip an exercise room, and employees think of that benefit as being much more than the $100,000 it cost the company, then the value of that investment to the company is much greater than

$100,000—and it's a good deal for the company and the employees. On the other hand, if the company spends $100,000 on a facility that employees don't use, and the employees think the company wasted the money, then that ill will might have a negative value to the company.

Small companies—or companies that are pressed for space or are located in high-rise office buildings—sometimes pay for their employees to join nearby health clubs. These supplements are subject to taxation, and employees may find the employer adding the cost of the membership to their paychecks. In that case, employees will pay tax on dollars the company spends for them to belong to the health club.

These tax implications are really only related to wellness programs that center around facilities. No one pays tax on programs that encourage employees to push the cafeteria tables back and do aerobics for twenty minutes after work. No one pays tax on CPR or first-aid classes. And the IRS won't come looking for a company that writes a smoking policy.

Let's Talk Legalese

The inevitability of taxes and their implications for wellness programs leads to businesses taking a look at the legal aspects of these programs as well. Many of the legal issues in worksite wellness programs are as yet untested in the courts. Businesses would be wise to consult their legal advisers.

Injuries to employees would most likely be handled through workers' compensation law. The issue gets less clear when employers pay for employees to join health and fitness centers away from the worksite. And the issue becomes even muddier when you consider other forms of injury, like second-hand smoke from a co-worker's cigarettes, or health problems arising from weight loss contests.

First, let's look at exercise programs. Who, if anyone, is responsible for employees who get hurt while doing aerobic exercise or playing a quick game of volleyball over the lunch hour or running in a corporate 10K? Put simply, if an employee is injured in an activity that is connected to employment, then the employee may seek a remedy in workers' compensation law. If, however, the injury does not grow out of employment, the employee may sue the employer in a civil action.

Trying to sort out what does and does not constitute employment is the key issue in legal liability when it comes to health promotion activities. Larson's *The Law of Workmen's Compensation* (1972) provides some basic principles. The following is a summary from section 22 of that work. Recreational or social activities are within the scope of employment when:

1. They occur on the premises during a lunch or recreation period as a regular incident of the employment; or
2. The employer, by expressly or impliedly requiring participation, or by making the activity part of the services of an employee, brings the activity within the orbit of the employment; or
3. The employer derives substantial direct benefit from the activity beyond the intangible value of improvement in employee health and morale that is common to all kinds of recreation and social life.

What circumstances can cause an activity to fall within the course of employment, even though the activity is not generally an activity for which the employee was hired? The following variables may apply:

1. Whether the activity occurred during working hours
2. Whether it was on the employer's premises
3. Whether participation was required
4. Whether the employer took the initiative in sponsoring or organizing the team
5. Whether the employer made contributions to the team
6. Whether the employer derived benefit from the team
7. Whether the employer directed the activity to take place
8. Whether the employer furnished equipment for the activity
9. Whether the employee expected compensation or reimbursement for the activity engaged in
10. Whether the activity was primarily for the personal enjoyment of the employee

Generally, the activity will be held to be within employment and therefore subject to the workers' compensation law, not civil law, if the answers to these questions seem to reveal employment-related activity.

Interestingly, the courts have held that improvement in employee health and morale are common to all kinds of recreation and social life, and in testing for whether the employer has derived substantial direct benefit from the activity, the court looks for more than improvement in employee morale and health.

If the employee is injured in a game played on the premises during a lunch or recreation period, the employee will generally be compensated for injuries. If the locale is off the premises but the players are allowed time off work for their games (company-sponsored), again, the employee may look to the workers' compensation court for compensation.

Larson points out that evidence of work connection (furnishing of financial support, athletic equipment, prizes, and the like by the employer) is helpful in a case of employer involvement, but, standing alone, it is ordinarily not enough to meet the burden of proof.

Entry forms for races usually include a waiver that the participant must sign in order to enter. A company could require its employees to sign a similar waiver before allowing them to use any facilities; however, an employer cannot disclaim all of its liability.

Not only physical injury but diseases are addressed under workers' compensation. If an employee attends a smoking cessation, alcohol treatment, or weight loss clinic, the following questions must first be answered:

1. Is the employer paying for the class?
2. Is the employee attending during working hours?
3. Does the employer benefit by anything more than the overall morale and physical fitness of the employee?
4. Is the program mandatory, or so highly suggested that it becomes mandatory?

A legal adviser offers the following fictitious examples: An employer is concerned with the obesity of an employee. The employer pays for and insists that the employee attend weight loss sessions during working hours. The weight loss information is poorly given, and the employee loses so much weight so fast that it affects some internal organs. The employee then incurs time off work for a disability and medical bills. That would be a workers' compensation case.

A second example: An employee attends a weight loss clinic, and the employer pays a portion of the tuition. Classes are held away from the worksite and not during working hours. Any liability growing from the loss of weight would not involve the employer.

A third example: The weight loss sessions are conducted by employees of the employer. Classes are voluntary, not held during working hours, and tuition is provided by the employee or on a shared basis with the employer. In this situation, injuries would probably not grow out of employment; however, if advice on weight loss was given and injuries resulted, the employee may be able to sue the employer in a civil case.

Employers, it seems, are exposed to a different breed of liability when it comes to worksite wellness programs, but all things considered, the benefits continue to outweigh the costs.

No Free Lunch

Health promotion programs don't exist in a vacuum within a company.

Companies often redesign the group health insurance plan to lower costs and tack on a health promotion program as an afterthought. Unfortunately, the direct effects of health promotion programs, when used in combination with other cost-cutting methods, are hard to measure. Shifting the cost of insurance and health care to employees may, in the short term, show quick results on the bottom line. But health promotion programs may prove to be the "magic bullet" in the long term.

Employers are using a potpourri of plans to try to do something about escalating health care costs. By working with insurance carriers, companies of all sizes are redesigning their medical plans. A smart executive will make sure that health promotion programs are part of the grand design, not just tacked on, or all of the realignments may be merely delaying the inevitable. Here's why.

A study by Hewitt Associates (1985), international consultants on employee benefits, showed that fewer than half of the companies surveyed felt their cost management efforts had had a significant or moderate impact on health care costs. These companies had used a number of methods, including shifting from first-dollar coverage to the use of deductibles, increasing the employee portion of coinsurance, making the employee pay for family coverage, auditing claims more closely, and offering financial incentives for employees to use outpatient surgery, preadmission testing, second-surgery opinions, home health care services, and to avoid nonemergency use of emergency rooms.

Shifting costs to employees puts more of the burden of being a watchdog on them and doesn't really make a difference in the overall cost of health care—somebody still pays for it. Certainly employers save money by increasing deductibles, and many companies are doing it. At the same time, companies, while shifting costs to employees, could upgrade other parts of the health care picture, especially by adding health promotion programs which include teaching employees how to be good health care consumers in the first place.

But the cost of the health care service—whether it's a day's stay in the hospital, a surgeon's fee for gall bladder surgery, or the cost of a lung x-ray—remains the same, no matter who pays. American consumers just don't shop around for bargains in health care, and they and their employers pay the retail price—no quibbling, no bargaining, no volume discounts. HMOs are an attempt to regulate the prices, but so far less than 10 percent of the population is using HMOs.

In the long run, cost shifting will accomplish little: "It would be like using a thimble to bail water out of a very leaky boat," according to Regina Herzlinger, professor of business administration at the Harvard Business School (1985). She blames increases in health care costs primarily on the mismatch between the demand for health care and the supply. The present system, she says, motivates health care providers to overproduce (build more hospitals) and consumers to overconsume (to take the attitude: "We have insurance, let's use it."). Again, Herzlinger puts it all in perspective when she likens the current system to "giving children open-ended credit cards to a candy factory."

The economic side of medicine isn't taught in medical school either. WELCOM's chairman, Dr. Eugene "Speedy" Zweiback, a surgeon and an avid runner, says, "Those of us who are on the firing line realize, of course, that cost-effective medicine is becoming as important as simple disease treatment was in the past. Certain potential conflicts arise since the physician has always viewed himself as the patient's advocate. I would suggest that that role need not change, but, at the same time, the physician must recognize a responsibility to the employer or provider who is obviously paying the bills" (personal letter to the author).

BUSINESS CAN'T WAIT FOR GOVERNMENT SOLUTIONS

Government interference in the free market system is something all business people moan and groan about, but government solutions aren't forthcoming anyway. And the ironic thing is that the people who are most adamant about government keeping its hands off the American market place are often the ones who are waiting for government to solve the problem of rising health care costs.

No one person, not even the President, and no political body, not even the Congress, can have an impact on health and its costs equal to that of the individual taking personal responsibility for his or her own well-being. The problem is not simply expensive hospitals. The problem is not our highly trained and highly paid doctors. The problem is us. On balance, too many Americans are afflicted with health problems that are directly related to their freely chosen lifestyle. The workplace is the logical point to break into the cycle of unhealthful lifestyles leading to poor health, higher health care costs, lost time at work, and finally to even more costly losses in productivity and life.

No federal dollars are available for health promotion in your office or factory. But in 1980 the Department of Health and Human Services outlined 227 health promotion objectives for the nation and gave the private sector 10 years to do something about those objectives (*Promoting Health/Preventing Disease: Objectives for the Nation*, 1980). The objectives for health promotion include the following specific targets for business and industry:

> *SMOKING:* In 1979, the proportion of the U.S. population that smoked was 33 percent. By 1990, the proportion of adults who smoke should be reduced to below 25 percent. Employers can en-

courage employees to stop smoking by providing smoking cessation programs and enforcing smoking policies in the workplace.

ALCOHOL AND DRUG MISUSE: Employers should provide substance abuse prevention and referral programs (employee assistance programs, EAP) for more than 70 percent of the workers in major firms. (Now about half of the Fortune 500 firms offer some type of employee assistance program.) Every employer, however, should consider writing a policy regarding the use of alcohol at company-related functions.

NUTRITION: By 1990, the prevalence of significantly overweight (120 percent of "desired" weight) people among the U.S. population should decrease to 10 percent of men (from 14 percent now) and 17 percent of women (from 24 percent now). Other objectives for improving nutrition also include reducing serum cholesterol levels and reducing the amount of salt people use. By posting calorie content and serving healthful foods in company cafeterias, employers can promote healthful diets.

EXERCISE: By 1990, the proportion of adults who participate in regular vigorous physical exercise should be greater than 60 percent (from around 35 percent now). On-site facilities for workouts are not always available or feasible for employees. But almost any worksite can become a locale for aerobics classes if management and employees work together to make it happen. By subsidizing memberships to health clubs and sponsoring employees who join recreational leagues for volleyball, bowling, and softball, the company can promote physical exercise without the expense of maintaining a facility.

STRESS: The government notes that 82 percent of people who took part in a recent study said they "need less stress in their lives." Through worksite health promotion programs, employers can do something about stress related to work and job satisfaction. The government report calls for more than 30 percent of the 500 largest companies in the United States to offer work-based stress reduction programs by 1990.

The Wellness Council of the Midlands—and its member companies—have adopted these objectives as their goal in promoting worksite wellness. And the SANE Approach to a Healthy Company in the areas of smoking, alcohol, nutrition, and exercise, outlined in chapter 3, grows directly from these goals.

These objectives are aimed toward the big companies, but the majority of the American work force is employed by medium-sized to small companies. The argument is made in this book that small and moderately sized companies are the ones that benefit most from membership in a Wellness Council. By taking advantage of the information-gathering ability of the council, small business can make its own sizable contribution to meeting these objectives.

The government always seems to be the bearer of bad tidings when it comes to health, and Surgeon Generals have been busy since 1957 telling us that cigarette smoking is bad for our health. We are told we aren't getting enough exercise, and so on. These government reports from task forces and committees and groups just keep reinforcing the fact that the government is reporting the trends; American business and the health care community must take the lead.

New programs are not the answer either, especially if they involve already hard-to-get federal dollars. In fact, in response to a report from the DHHS that blacks and other minorities are less healthy and die younger than whites, former Secretary Margaret Heckler said, "I do not believe that money is the answer. . . . Progress depends more on education and a change in personal behavior than it does on more doctors, more hospitals or more technology" (*Omaha World-Herald*, October 1985).

How does the government expect the American public to meet its self-imposed health objectives in the next few years if the government doesn't kick in some money to do it? Just because the federal government makes the case for healthy people, the effort in carrying out the government's mandates must be collective and have local roots. A single company or a Wellness Council of companies can meet—and beat—the objectives for the nation for 1990.

President Reagan agrees. He told WELCOM, "What you are doing in the Wellness Council responds to this Administration's belief that private initiatives rather than government programs can be of great benefit to the nation in dealing with many social and community needs" (personal letter to the author).

CORPORATE CONCERN FOR QUALITY FAMILY LIFE

Worksite wellness can create a strong role for corporate America in promoting the quality of family life for its workers. As powerful a vehicle as the workplace is in promoting wellness, no place has the potential for long-term impact on the health of America that families do. The workplace can in fact become a pipeline of information and behavior change into the family.

Working with employees in the workplace is one thing, but broadening the issue to concern for families is another. Is it even fair, much less realistic, for the private sector to get involved in the quality of their employees' family lives?

The answer is a definite "yes," for a variety of good reasons. The first is that, like it or not, no other institution in American life, year in and year out, has an impact on the family like that of the employer(s) of a family's breadwinner(s). In a democratic society, one's job in large measure determines the social status of a family. It determines peers and friendship patterns, and often determines in what neighborhood (perhaps even in what city) a family will reside, and thus determines where the children will go to school, where the family will do its primary shopping, and may even determine which church the family will attend.

The impact of the workplace goes well beyond a family's socioeconomic status. Jobs determine a whole array of benefits such as life insurance, health, dental, and eye care, pension programs, and profit sharing. All of these have a dramatic impact on the quality of family life. Now put health promotion programs into that equation and, just as many companies have done with employee assistance programs, make them available to family members of employees. Companies generally make medical and dental benefits available to employees' family members; health promotion efforts can also be designed to include spouses and children.

No Better Place Than Work

Another reason for companies to address family issues is that no other institution can do it as well. The workplace has a golden opportunity, in light of government cutbacks to social service agencies, to demonstrate the capacity of the private, for-profit sector to tackle major social problems. The workplace can disseminate information more effectively and more efficiently than any other institution because so many people are now there, especially for blacks and other minorities, despite their high levels of unemployment.

Beyond the function of distributing information, the workplace has the opportunity to ease the tension on families through simple changes in policy and procedures. The workplace, through quality training and information programs, can help to shape new values and new ways of balancing careers and personal values. In fact, the corporation may be the last institution able to shape and mold new values. I say this because the tremendous mobility of our society and the state of upheaval of public education, churches, public and private social service agencies, and the upheaval within families themselves make these traditional means of reaching people less effective.

The Bottom Line—Product and Productivity

Private-sector employers may want to become involved in helping the American family adjust to this new age for reasons that affect the bottom line. A middle manager may be a company's shining star, but if he is living in constant disharmony at home, or if his teenage daughter, whom he suspects is using street drugs, did not come home until 3 a.m. last Saturday night, this promising manager is not going to be particularly efficient at even routine daily tasks. A worksite wellness program could help.

An ongoing series of family life seminars, for example, could be offered during after-work hours or as brown-bag lunch seminars. Many companies are offering a variety of programs on such topics as communication for couples, dual careers, single parenting, disciplining children, living with adolescents, family money management, and simple home maintenance. Informed employees who have some good information and a little training can avoid the kinds of problems that often lead to more expensive responses (employee assistance programs).

Here's another scenario: A woman may arrive at work 10 minutes late each day because she is part of a carpool that takes her kids to school. Her supervisor has issued warnings: The woman is experiencing a lot of stress but feels trapped, unable to change the driving arrangements. A health promotion program that allows for flexible hours would not only enable this woman to work the required amount of time each week, but might also help her to reduce stress by encouraging her to attend an aerobics class two or three times each week.

As more and more women work, the workplace has had to make adjustments for the concerns of women. This has often been done reluctantly, and sometimes even when business vehemently did not want to do so. The surge of mothers into offices, stores, and factories has brought with it increasing pressure for flexible work hours, job sharing, and child care facilities—or at least the freedom to arrange work around family. Worksite wellness efforts, in turn, benefited from the influx of working mothers. Women, having forced the workplace to change and respond in new ways to their personal concerns, paved the way for acceptance of worksite wellness activities.

Many people do not really understand just how dramatic the movement of women into the work force has been. When examined, the numbers are staggering and help us understand why the influx of women into the workplace is one of the most important social changes of the twentieth century.

Figures from the Bureau of Labor Statistics (BLS) help us understand the phenomenon. In 1947, 18 percent of women with children worked. In 1980, the figure had risen to 60 percent. The BLS predicts that in 1990 the number of working women with children will exceed 70 percent.

Progressive companies are also discovering that there is a creative and energetic mix when competent and well-trained men and women work together in an egalitarian environment. More and more often, the job of the manager is not getting a job done, but getting other people to get many jobs done. In such an environment, the most important leadership traits are those traditionally, perhaps culturally, best demonstrated by women: enabler, nurturer, teacher, helper, problem solver, small-group leader, mentor, and counselor. Progressive companies understand the value of these people-oriented traits, especially when combined with good education and training.

These same traits, which reveal a holistic view of people, create the kind of environment in which a worksite wellness program can flourish. The idea of helping people to take personal responsibility for their total well-being is not so foreign to the astute woman manager as it often is to their sometimes slow-to-learn male counterparts.

Corporate involvement in the quality of family life goes beyond even the bottom line. It addresses the issue of the survival of the American way of life because the values that make a free capitalist society work are taught and nurtured in healthy family environments. American business and industry are reshuffling their priorities—especially during recessions and market downturns. They are keeping an eye on the bottom line, but at the same time are experimenting with flexible hours, split shifts, job sharing, and putting computer terminals in private homes.

As an integral part of the growing worksite wellness movement, corporate concern for family is the preventive and much less costly side of employee assistance programs. In the short term, worksite wellness programs show an impact on the morale of employees and on productivity. In the long run, health promotion has the potential to enhance the best aspects of the American way of life, year after year, in the American family—and to affect the bottom line.

2

A BUSINESS PLAN FOR HEALTH PROMOTION AT WORK

Wellness programs have traditionally been introduced into a corporate setting with little or no thought. Naturally, this runs counter to the business tradition of short- and long-range planning. A business person would not hire a new employee without knowing what his or her responsibilities would be, where that person would fit within the corporate structure, or whether the company had enough capital to justify a new salary line.

That same decision maker would not order a mainframe computer system without fully researching its use, its costs, where to place it, and what support equipment is needed. Yet that methodical decision maker might decide one day that the employees need a weight loss program, arrange to have it provided, and wonder why nobody showed up.

As with any other aspect of business, a worksite wellness program needs to go through a planning process. In fact, the plan is the most important element. The business plan presented in this book is easy to follow and leads decision makers toward a wellness program that does not cost a lot in money and personnel time, but has the potential to pay off on any company's bottom line.

Figure 2.1 illustrates the steps of this business plan, and each step is discussed in detail in this chapter. The SANE Approach to a Healthy Company—shown in the figure as a branch of strategic planning—offers any company four specific things to do right away, with little commitment of money and not much personnel time. These involve the areas of smoking, alcohol, nutrition, and exercise. This quick route to worksite wellness sets the stage for more detailed, more expensive

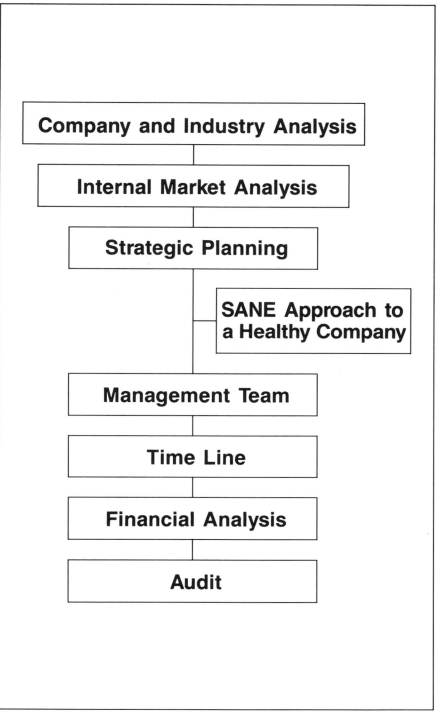

Figure 2.1. Outline of a business plan for health promotion at work.

health promotion programs. The SANE approach is fully explored in the next chapter.

Like any major corporate undertaking, it is best to launch a good wellness program with an organized, businesslike approach. This chapter develops a step-by-step business plan for starting, implementing, and evaluating a wellness program. The basic principles of business planning apply to the development of workplace health promotion programs.

The format, as with any good business plan, helps decision makers first to ask probing questions and then to project realistic answers. In the section on COMPANY AND INDUSTRY ANALYSIS, business leaders are asked to consider what business they are in and to look at the external and internal competitors. Then it is appropriate to examine the makeup of the work force and ask employees what programs they need and what programs they would participate in. The section on INTERNAL MARKET ANALYSIS presents several survey forms that any company can use.

The step-by-step plan should prove especially helpful to decision makers in companies where little has been done in the area of health promotion so far. Because the information that follows is built around basic planning principles, it will help to design wellness goals and objectives that fit into the overall corporate plan. The section on STRATEGIC PLANNING covers those long- and short-term goals, discusses the risks of doing nothing, and looks at why businesses initiate worksite wellness programs.

The section on MANAGEMENT TEAM introduces the idea of employee ownership. It's a key strategy and well worth exploring. Even though the impetus for worksite programs comes best from the top of a company, keeping middle management's support is difficult. Some techniques are discussed.

By setting dates and durations for programs, the company can control the TIME LINE.

What do these programs cost? And what is sickness costing the company now? That is discussed in the section on FINANCIAL ANALYSIS, along with a reasonable budget and realistic returns on the investment in human capital.

How do you know if the programs are working? The AUDIT shows simple approaches to evaluating programs and their effectiveness.

For companies that clearly have some level of wellness activities, this business plan will help them refine their planning process. More important, perhaps, it will offer how-to insights into improving employee participation, getting management support, evaluating programs, and measuring cost effectiveness and the impact on employee health.

As the plan is laid out over the following pages, it goes from the basic, low-cost programs to sophisticated and expensive ones. Business leaders will probably be surprised to learn how much can be done at little or no cost. Whether a company begins with low-cost programs or at any other place on the spectrum, this chapter should be helpful.

In any company, worksite wellness runs the risk of becoming a passing fad, a temporary experiment, an executive perk, or a phased-out fluff program. Worksite wellness, if it is to become an integral part of the corporate culture—in fact, to shape a healthy company—must take its place in the planning cycle and be held equally accountable with all other departments. Perhaps in the early stages at least, it must be among a company's most professionally planned and promoted activities. The following business plan should help those businesses, large and small, who are serious about worksite wellness.

COMPANY AND INDUSTRY ANALYSIS

Every company is in the "people" business, whether it is a one-person operation or employs 100,000 workers. The design, sales, and distribution of products and services cannot be accomplished without people.

The goals and objectives of each business are directly related to the productivity of these people or employees. The old cliché that a chain is only as strong as its weakest link is apropos here. For a company to achieve high productivity, employees must, first of all, be at work. Then, while at work, they must be able individually to perform well to be productive. A sick or absent employee, therefore, will decrease a company's level of productivity.

In a real sense, the absent and sick employees become competitors to the business they work for. With decreased productivity comes payment for sick days, medical and hospitalization costs, and the future possibility of increased insurance payments based on higher levels of use (for example, if you have car accidents, your car insurance rates will rise). The dilemma is therefore two-sided: A lower level of productivity means lower profits, on one side, and increased expenditures for insurance and sick leave mean lower profits, on the other side. Either way, the company loses.

Competition is keen enough in the market place without having to fight internal competitors. Just as they do in competitive situations, businesses have options for action. In the market place it may include increasing the advertising budget, exploring new markets, and developing more product lines. Internally, it may include redesigning the benefits package, exploring alternative health care options, and implementing health promotion programs at the worksite.

"Doing nothing" is not an option in any of these situations. Rising health care costs will devour the profits. The livelihood of the business is at stake. Businesses did nothing for years as healthy profits paid the insurance premiums and supported generous benefit packages. This, along with an attitude of laissez faire toward the health of employees, has brought American business to its present crossroads.

For some corporations the trend is beginning to shift. They are finding that putting an emphasis on the employee (for example, by starting health promotion programs) is paying off on the bottom line. The investment in human capital, a company's most important asset, is undoubtedly the best investment a company can make. It's like buying a service maintenance contract on employees. These corporations realize that the real competitors in the market place will be progressive companies that have implemented the necessary changes to remain on the cutting edge of their respective businesses—whether they manufacture something or sell a service.

To get a good start, each business leader needs to make a company and industry analysis. The following questions need to be answered:

What business are we really in?

Who are our external competitors (other companies), and what are our internal competitors (high rates of sick leave, turnover, burnout, accidents)?

What are other companies in our industry doing? Is our business cyclical, seasonal, facing a downturn or an upswing? And is the work force physically and mentally prepared to meet those challenges?

What is generally being done in corporate America with health promotion at the worksite?

Which local businesses are active in worksite wellness? Can we join a local Wellness Council and benefit from the experiences of other businesses in our community, while contributing to the overall wellness of the community in which we do business and from which we hire our work force?

INTERNAL MARKET ANALYSIS

Closely related to company and industry analysis is the internal market analysis. The background is established by addressing the previous questions. The next step is to compile a detailed analysis of your market—that is, your employees. The basic demographics of age, gender, and number and kinds of family members can be obtained through various internal channels like the personnel files or human resources data.

The configuration of the company itself will make a difference in what health promotion programs the company can do. How many locations does the company have in the community? In the region? Nationally? How much territory do the employees cover during a workday? For example, if there are many branch offices throughout the country, corporate smoking and alcohol use policies can work as an extension of the main philosophy of the company. Or if there is a central headquarters, but many employees work in the field, as in the phone company or electrical and gas utilities, then on-site workshops may be difficult to schedule and may not be well attended by that segment of the work force.

If a company hires a great number of minimum-wage employees who work part-time and on swing shifts, the best approach for them might be seminars and exercise classes just before or after the shifts.

Sometimes the wellness program can be the springboard for action programs. At Union Pacific Railroad, for example, the Brotherhood of Locomotive Engineers and the United Transportation Union have joined with the Union Pacific System throughout the country in a program called Operation Red Block. It's a revolutionary five-step drug and alcohol prevention program in which labor and management work together to create an environment that is free of the risks induced by alcohol and drugs. Using the theme, "Your Right to Get High Ends Where My Shift Begins," prevention committees formed in the local unions field complaints about members using drugs and alcohol while on duty. Committees insist that users quit and urge them to contact the employee assistance counselor if they need help.

Programs such as Operation Red Block go a long way to improve working conditions and to address the concerns employees have. To find out what employees want and need, employers may use a number of survey forms that are already available. Employee interests can be found out by means of an employee interest survey. Employee needs can be obtained through a needs assessment instrument or health risk appraisal. These various tools are discussed below.

Finding Out What Employees Want

Companies can design their own interest surveys, or they can use either of the surveys presented here. How detailed the instrument needs to be depends entirely on what type of information the designer wants. The Program Interest Questionnaire in figure 2.2 is simple but asks for fundamental information; whereas, the Employee Wellness Survey in figure 2.3 is more sophisticated and requires more time for the employee to complete and the company to analyze. Another option is to redesign one of these questionnaires by starting with a simple structure and adding questions the company needs answers for.

Program Interest Questionnaire

Program Interests: The following are examples of the types of programs that might be offered. Please indicate by circling the appropriate response how interested you would be in participating.

	Very Interested	Somewhat Interested	Somewhat Uninterested	Not Interested
An introductory program that explains how you can become all that you can be and avoid disease, and which would help you to design and plan your own wellness program	1	2	3	4
A stop-smoking program	1	2	3	4
A program on nutrition	1	2	3	4
A program on stress management	1	2	3	4
A program on exercise for good health to help you begin a personalized exercise program	1	2	3	4
A program on weight reduction	1	2	3	4
A dancing-for-fitness program	1	2	3	4
A program on accident prevention and safety	1	2	3	4
A communications program to help you improve your human relationships	1	2	3	4
A recreation program (such as softball, basketball, volleyball)	1	2	3	4
If there was an opportunity to do so, would you want your family to participate in these programs?	1	2	3	4

Figure 2.2. Program interest questionnaire. Used with permission from Possibilities, Inc.

Employee Wellness Survey

Please take a few moments to fill out. From the following list of activities and programs, check the ones you would be most interested in and if you would attend them.

Interested   **Will Attend**

Interested	Will Attend		
_____	_____	1.	Alcohol and Drug Abuse Control
_____	_____	2.	Assertiveness Training
_____	_____	3.	Cancer Risk Reduction
_____	_____	4.	Cardiopulmonary Resuscitation (CPR)
_____	_____	5.	Positive Thinking
_____	_____	6.	Dental Disease Prevention
_____	_____	7.	First Aid
_____	_____	8.	Health Awareness Program
_____	_____	9.	Glaucoma Screening
_____	_____	10.	Goal Achievement
_____	_____	11.	Healthy Back
_____	_____	12.	Heart Attack Risk Reduction
_____	_____	13.	Hypertension Control
_____	_____	14.	Nutrition
_____	_____	15.	Physical Fitness Activities
_____	_____	16.	Safety-Accident Prevention
_____	_____	17.	Stress Management
_____	_____	18.	Weight Management
_____	_____	19.	Physical Fitness Lectures
_____	_____	20.	Other _____

2. Would you attend one or more of the above programs if they were offered at a convenient time?
___Yes ___No ___Maybe
What time is most convenient for you? _____

3. Do you have a planned, regular program of exercise (swimming, walking, jogging, exercise machines) in which you participate at least three times a week? ___Yes ___No

4. Would you like to participate in an exercise/fitness program that was geared to your level of fitness?
___Yes ___No

5. What activities would you like to learn?
_____
_____
_____

6. What types of exercise programs would you like to see at the worksite?
_____ Aerobic
_____ Jazzercize
_____ Calisthenics
_____ Other _____

7. Would your spouse and/or family participate in a health promotion program at your worksite if they were invited? ___Yes ___No

8. If one or more of the programs or activities listed above that you selected as the most interesting to attend were offered at a convenient time, and at a reasonable cost, would you probably attend that program? ___Yes ___No

9. Would you feel comfortable participating in a program with your co-workers? ___Yes ___No

10. In the space below, write any other health care or health promotion ideas or concerns that may have been triggered by the previous questions.

What hours do you work?_____ a.m./p.m. to _____ a.m./p.m.
Are you _____ Male or _____ Female?
What age group are you in? _____ 20-30
                                            _____ 31-40
                                            _____ 41-50
                                            _____ Over 50
Are you: _____ Administrative or ___ Shop?

Figure 2.3. Employee wellness survey. Adapted with permission from Valmont Industries, Inc., Valley, Nebr.

45

Finding Out What Employees Need

Needs assessment instruments or health risk appraisals focus more on the health habits of the employees than on their interests. Some of the information may include how many smokers there are in the company, how many people regularly exercise, and how many use seat belts. The more sophisticated instruments ask for blood pressure, blood analysis, and other more detailed health statistics.

A note of caution: The less sophisticated surveys are good for gathering general information about the people who work for the company. They also act as an awareness tool for the individual participants. They can be the first step in conditioning the corporate climate for upcoming wellness programs.

Simple self-scoring health risk appraisals can be the first step in plowing the corporate field prior to the cultivation of good wellness programs. It is not necessary to have the results individually interpreted; however, most health risk appraisals go beyond information and awareness. These instruments can be helpful with individual behavioral change patterns when the results are distributed and interpreted correctly. In-house and community resources must also be available so individuals can be referred for assistance. (For example, a health risk appraisal might uncover someone who wants to quit smoking but doesn't know how. The person could be referred to a smoking cessation program conducted by the American Lung Association or the American Cancer Society.)

A basic needs assessment is all that is required for the initial market analysis. The objective is to learn about the market (the employees), not to gear up for programming for them—not yet. The needs assessment operates on the standard principle that you can only please some of the people some of the time.

About 20 percent of the employees in any company will generally be fit. They eat right, probably exercise, and watch their intake of fats, salt, and sugar. A large group (about 60 percent) will be "near well." Some of these people are the smokers; some are overweight; many probably don't exercise. This group could prolong their lives, be more productive workers, use their health insurance less often, decrease sick days, and be healthier and happier if they made lifestyle changes. These are the people at whom the health promotion programs are aimed.

The remaining 20 percent are people who are at risk for serious disease because of their family history or because of their own destructive lifestyles (heavy smokers, always stressed, seriously overweight). Health promotion programs would be great for them, but these at-risk employees are not the direct target of wellness programming.

Figure 2.4 shows a simple but effective instrument called Health-

All of us want good health. But many of us do not know how to be as healthy as possible. Health experts now describe *lifestyle* as one of the most important factors affecting health. In fact, it is estimated that as many as seven of the ten leading causes of death could be reduced through common-sense changes in lifestyle. That's what this brief test, developed by the Public Health Service, is all about. Its purpose is simply to tell you how well you are doing to stay healthy. The behaviors covered in the test are recommended for most Americans. Some of them may not apply to persons with certain chronic diseases or handicaps, or to pregnant women. Such persons may require special instructions from their physicians.

Healthstyle
A Self-Test

	Almost Always	Sometimes	Almost Never
Cigarette Smoking			

If you *never smoke*, enter a score of 10 for this section and go to the next section on *Alcohol and Drugs*.

	Almost Always	Sometimes	Almost Never
1. I avoid smoking cigarettes.	2	1	0
2. I smoke only low tar and nicotine cigarettes *or* I smoke a pipe or cigars.	2	1	0

Smoking Score: _____

Alcohol and Drugs

	Almost Always	Sometimes	Almost Never
1. I avoid drinking alcoholic beverages *or* I drink no more than 1 or 2 drinks a day.	4	1	0
2. I avoid using alcohol or other drugs (especially illegal drugs) as a way of handling stressful situations or the problems in my life.	2	1	0
3. I am careful not to drink alcohol when taking certain medicines (for example, medicine for sleeping, pain, colds, and allergies), or when pregnant.	2	1	0
4. I read and follow the label directions when using prescribed and over-the-counter drugs.	2	1	0

Alcohol and Drugs Score: _____

Eating Habits

	Almost Always	Sometimes	Almost Never
1. I eat a variety of foods each day, such as fruits and vegetables, whole grain breads, and cereals, lean meats, dairy products, dry peas and beans, and nuts and seeds.	4	1	0
2. I limit the amount of fat, saturated fat, and cholesterol I eat (including fat on meats, eggs, butter, cream, shortenings, and organ meats such as liver).	2	1	0
3. I limit the amount of salt I eat by cooking with only small amounts, not adding salt at the table, and avoiding salty snacks.	2	1	0
4. I avoid eating too much sugar (especially frequent snacks of sticky candy or soft drinks).	2	1	0

Eating Habits Score: _____

Exercise/Fitness

	Almost Always	Sometimes	Almost Never
1. I maintain a desired weight, avoiding overweight and underweight.	3	1	0
2. I do vigorous exercises for 15-30 minutes at least 3 times a week (examples include running, swimming, brisk walking).	3	1	0
3. I do exercises that enhance my muscle tone for 15-30 minutes at least 3 times a week (examples include yoga and calisthenics).	2	1	0
4. I use part of my leisure time participating in individual, family, or team activities that increase my level of fitness (such as gardening, bowling, golf, and baseball).	2	1	0

Exercise/Fitness Score: _____

Figure 2.4. Healthstyle. From *Healthstyle: A Self-Test,* National Health Information Clearinghouse, Washington, D.C.

	Almost Always	Sometimes	Almost Never
Stress Control			
1. I have a job or do other work that I enjoy.	2	1	0
2. I find it easy to relax and express my feelings freely.	2	1	0
3. I recognize early, and prepare for, events or situations likely to be stressful for me.	2	1	0
4. I have close friends, relatives, or others whom I can talk to about personal matters and call on for help when needed.	2	1	0
5. I participate in group activities (such as church and community organizations) or hobbies that I enjoy.	2	1	0

Stress Control Score: _____

	Almost Always	Sometimes	Almost Never
Safety			
1. I wear a seat belt while riding in a car.	2	1	0
2. I avoid driving while under the influence of alcohol and other drugs.	2	1	0
3. I obey traffic rules and the speed limit when driving.	2	1	0
4. I am careful when using potentially harmful products or substances (such as household cleaners, poisons, and electrical devices).	2	1	0
5. I avoid smoking in bed.	2	1	0

Safety Score: _____

What Your Scores Mean to YOU

Scores of 9 and 10
Excellent! Your answers show that you are aware of the importance of this area to your health. More important, you are putting your knowledge to work for you by practicing good health habits. As long as you continue to do so, this area should not pose a serious health risk. It's likely that you are setting an example for your family and friends to follow. Since you got a very high test score on this part of the test, you may want to consider other areas where your scores indicate room for improvement.

Scores of 6 to 8
Your health practices in this area are good, but there is room for improvement. Look again at the items you answered with a "Sometimes" or "Almost Never." What changes can you make to improve your score? Even a small change can often help you achieve better health.

Scores of 3 to 5
Your health risks are showing! Would you like more information about the risks you are facing and about why it is important for you to change these behaviors. Perhaps you need help in deciding how to successfully make the changes you desire. In either case, help is available.

Scores of 0 to 2
Obviously, you were concerned enough about your health to take the test, but your answers show that you may be taking serious and unnecessary risks with your health. Perhaps you are not aware of the risks and what to do about them. You can easily get the information and help you need to improve, if you wish. The next step is up to you.

Figure 2.4. Continued

style, which is provided by the National Health Information Clearinghouse. Confidentiality can be maintained by asking the participants not to write their names on the questionnaire. Awareness of what you need to do to change your lifestyle is half the battle in worksite wellness. By scoring the questionnaire on their own, employees can test themselves and know on the spot what areas they need to improve.

To insure greater participation, give the surveys to groups of em-

YOU Can Start Right Now!

In the test you just completed were numerous suggestions to help you reduce your risk of disease and premature death. Here are some of the most significant:

Avoid cigarettes. Cigarette smoking is the single most important preventable cause of illness and early death. It is especially risky for pregnant women and their unborn babies. Persons who stop smoking reduce their risk of getting heart disease and cancer. So if you're a cigarette smoker, think twice about lighting that next cigarette. If you choose to continue smoking, try decreasing the number of cigarettes you smoke and switching to a low tar and nicotine brand.

Follow sensible drinking habits. Alcohol produces changes in mood and behavior. Most people who drink are able to control their intake of alcohol and to avoid undesired, and often harmful, effects. Heavy, regular use of alcohol can lead to cirrhosis of the liver, a leading cause of death. Also, statistics clearly show that mixing drinking and driving is often the cause of fatal or crippling accidents. So if you drink, do it wisely and in moderation.

Use care in taking drugs. Today's greater use of drugs—both legal and illegal—is one of our most serious health risks. Even some drugs prescribed by your doctor can be dangerous if taken when drinking alcohol or before driving. Excessive or continued use of tranquilizers (or "pep pills") can cause physical and mental problems. Using or experimenting with illicit drugs such as marijuana, heroin, cocaine, and PCP may lead to a number of damaging effects or even death.

Eat sensibly. Overweight individuals are at greater risk for diabetes, gall bladder disease, and high blood pressure. So it makes good sense to maintain proper weight. But good eating habits also mean holding down the amount of fat (especially saturated fat), cholesterol, sugar, and salt in your diet. If you must snack, try nibbling on fresh fruits and vegetables. You'll feel better—and look better, too.

Exercise regularly. Almost everyone can benefit from exercise—and there's some form of exercise almost everyone can do. (If you have any doubt, check first with your doctor.) Usually, as little as 15–30 minutes of vigorous exercise three times a week will help you have a healthier heart, eliminate excess weight, tone up sagging muscles, and sleep better. Think how much difference all these improvements could make in the way you feel!

Learn to handle stress. Stress is a normal part of living; everyone faces it to some degree. The causes of stress can be good or bad, desirable or undesirable (such as a promotion on the job or the loss of a spouse). Properly handled, stress need not be a problem. But unhealthy responses to stress—such as driving too fast or erratically, drinking too much, or prolonged anger or grief—can cause a variety of physical and mental problems. Even on a very busy day, find a few minutes to slow down and relax. Talking over a problem with someone you trust can often help you find a satisfactory solution. Learn to distinguish between things that are "worth fighting about" and things that are less important.

Be safety conscious. Think "safety first" at home, at work, at school, at play, and on the highway. Buckle seat belts and obey traffic rules. Keep poisons and weapons out of the reach of children, and keep emergency numbers by your telephone. When the unexpected happens, you'll be prepared.

Figure 2.4. Continued

ployees at regular employee meetings. Such meetings can also be an opportunity to educate employees on risk factors and lifestyle. Such meetings can also generate (or regenerate) excitement for company wellness programs.

A needs assessment might show that the company has a lot of smokers. The employee interest survey might show that most of the smokers want to stop smoking. BINGO! The combination of needs and wants leads directly into a smoking cessation program, and it is likely that

Where Do You Go From Here:

Start by asking yourself a few frank questions: Am I really doing all I can to be as healthy as possible? What steps can I take to feel better? Am I willing to begin now? If you scored low in one or more sections of the test, decide what changes you want to make for improvement. You might pick that aspect of your lifestyle where you feel you have the best chance for success and tackle that one first. Once you have improved your score there, go on to other areas.

If you already have tried to change your health habits (to stop smoking or exercise regularly, for example), don't be discouraged if you haven't yet succeeded. The difficulty you have encountered may be due to influences you've never really thought about—such as advertising—or to a lack of support and encouragement. Understanding these influences is an important step toward changing the way they affect you.

There's Help Available. In addition to personal actions you can take on your own, there are community programs and groups (such as the YMCA or the local chapter of the American Heart Association) that can assist you and your family to make the changes you want to make. If you want to know more about these groups or about health risks, contact your local health department or mail in the coupon below. There's a lot you can do to stay healthy or to improve your heath—and there are organizations that can help you. Start a new HEALTHSTYLE today!

Figure 2.4. Continued

employees will participate and be successful. Whereas if the needs assessment showed that many people were overweight, and the interest survey showed that employees would rather have information on CPR or first aid classes, then a weight-loss contest might not go over well. It also shows that work must be done to create an incentive to lose weight.

A health risk appraisal, provided by the Centers for Disease Control (CDC) in Atlanta, shown here as figure 2.5, can be used for information and awareness. However, health risk appraisals are more useful during the implementation of the worksite wellness program.

Compiling health care costs is another key component of the market analysis. A company needs to look at what health care is costing the company and where the problem areas are. Perhaps the human resources director suspects that employees are spending too many days in the hospital for routine surgery. Maybe there is a rise in foot ailments because a number of podiatrists have begun practice in the area. What specific areas are traditionally high for the particular company? Back problems, for example, might plague manufacturing companies. What traditional illness incentives are provided to the employee? Do employees use all the sick leave they are allowed on a "use it or lose it" arrangement? Or are they paid for unused sick leave? If deductibles are low, do employees seem to go to the doctor more often? Some of this information can be obtained from company records, but some will have to be sought directly from the insurance carriers. Carriers used to resist requests by businesses for data on medical claims. Companies didn't really know what was spent for medical claims, nor did they care.

Now it's important to know where the health care dollar is being spent. In fact, insurance carriers are generally eager to provide data on health care expenditures as a part of the cost-containment features they provide to their client companies. It's not so important anymore how long it takes the insurance company to process billing; businesses now look to their insurance carriers to provide information and educational services as well.

Companies are putting their insurance claims processing up for bid these days. It used to be that a company would choose an insurance company and stay with that company forever—never questioning the processing and practices, never asking for information. But that has all changed. Insurance carriers are becoming increasingly more responsive to the needs of client companies.

A final area to check in the internal market analysis is the corporate culture—in other words, "the way things are done around here." Companies far too often pay little or no attention to this crucial area. Individual change as well as collective corporate change is much more difficult to bring about if the corporate environment does not nurture change. Corporate culture is an expansive—not necessarily expensive—area but it needs to be addressed. It ranges from corporate alcohol policies, to the installation of smoke-removing machines in designated smoking areas, to unwritten but established office procedures. The objective in creating a healthy corporate culture is to identify the unhealthy areas of the work environment and determine which can be changed.

STRATEGIC PLANNING

Once the need has been established and the various analyses have been made, the next phase is to formulate a plan. This is when the general goals of the worksite program should be established. One common goal is to create a healthy environment in which a healthier, more productive, more loyal employee is absent less often and has a higher level of morale. Associated with these attributes are increased productivity, lower health care costs, and a greater competitive edge in the market place.

One study of small businesses found that health care cost savings were not as important to the companies surveyed as enhancing morale, reducing absenteeism, and improving employees' overall health and well-being.

Along with the goals, expectations need to be addressed. What are the expectations of management, or what can feasibly be accomplished in one month, six months, a year, two years, and longer? What do the decision makers expect will be accomplished in a set period of time?

(Detach Here)

(B) HEALTH RISK APPRAISAL

Health Risk Appraisal is a promising health education tool that is still in the early stages of development. It is designed to show how your individual lifestyle affects your chances of avoiding the most common causes of death for a person of your age, race and sex. It also shows how much you can improve your chances by changing your harmful habits. (This particular version is not very useful for persons under 25 or over 60 years old and for persons who have had a heart attack or other serious medical problem.)

IMPORTANT: To assure protection of your privacy, do NOT put your name on this form. Make sure that you put your Health Risk Appraisal "coupon" in your wallet or other safe place and insure that the number matches the number on this form. You must present your coupon to get your computer results.

PARTICIPANT NUMBER [_____]   1-6

PLEASE ENTER YOUR ANSWERS IN THE EMPTY BOXES (use numbers only)

1. SEX   [1] Male   [2] Female   7

2. RACE/ORIGIN
[1] White (non-Hispanic origin)   [2] Black (non-Hispanic origin)   [3] Hispanic
[4] Asian or Pacific Islander   [5] American Indian or Alaskan Native   [6] Not sure   8

3. AGE (At Last Birthday)   Years Old   9-10

4. HEIGHT (Without Shoes)   Example: 5 foot, 7½ inches = [5] ' [0][8] "   (No Fractions)   11-13

5. WEIGHT (Without Shoes)   Pounds   14-16

6. TOBACCO   [1] Smoker   [2] Ex-Smoker   [3] Never Smoked   17

(Smokers and Ex-smokers) — Enter average number smoked per day in the last five years (ex-smokers should use the last five years before quitting.)
- Cigarettes Per Day   18-19
- Pipes/Cigars Per Day (Smoke Inhaled)   20-21
- Pipes/Cigars Per Day (Smoke Not Inhaled)   22-23

(Ex-smokers only)   Enter Number of Years Stopped Smoking (Note: Enter 1 for less than one year)   24-25

7. ALCOHOL   [1] Drinker   [2] Ex-Drinker (Stopped)   [3] Non-Drinker (or drinks less than one drink per week)   26

If you drink alcohol , enter the average number of drinks per week:
- Bottles of beer per week   27-28
- Glasses of wine per week   29-30
- Mixed drinks or shots of liquor per week   31-32

8. DRUGS/MEDICATION How often do you use drugs or medication which affect your mood or help you to relax?
[1] Almost every day   [2] Sometimes   [3] Rarely or Never   33

9. MILES Per Year as a driver of a motor vehicle and/or passenger of an automobile (10,000 = average)   Thousands of miles   [0][0][0]   34-38

10. SEAT BELT USE (percent of time used)   Example: about half the time = [5][0]   %   39-41

11. PHYSICAL ACTIVITY LEVEL
[1] Level 1 - little or no physical activity
[2] Level 2 - occasional physical activity
[3] Level 3 - regular physical activity at least 3 times per week   42

NOTE: Physical activity includes work and leisure activities that require sustained physical exertion such as walking briskly, running, lifting and carrying.

12. Did either of your parents die of a heart attack before age 60?
[1] Yes, One of them   [2] Yes, Both of them   [3] No   [4] Not sure   43

13. Did your mother, father, sister or brother have diabetes?   [1] Yes   [2] No   [3] Not sure   44

14. Do YOU have diabetes?   [1] Yes, not controlled   [2] Yes, controlled   [3] No   [4] Not sure   45

15. Rectal problems (other than piles or hemorrhoids).
Have you had:
- Rectal Growth?   [1] Yes   [2] No   [3] Not sure   46
- Rectal Bleeding?   [1] Yes   [2] No   [3] Not sure   47
- Annual Rectal Exam?   [1] Yes   [2] No   [3] Not sure   48

CDC 90.4 4-82   (Continued on Other Side)

Figure 2.5. Health risk appraisal. Source: Centers for Disease Control, Atlanta, Ga.

16. Has your physician ever said you have Chronic Bronchitis or Emphysema?　　　① Yes　　　② No　　　③ Not sure　　　□ 49

17. Blood Pressure (If known – otherwise leave blank)　　　Systolic (High Number)　　　□□□ 50-52
　　　Diastolic (Low Number)　　　□□□ 53-55

18. Fasting Cholesterol Level (If known – otherwise leave blank)　　　MG/DL　　　□□□ 56-58

19. Considering your age, how would you describe your overall physical health?
　　　① Excellent　　　② Good　　　③ Fair　　　④ Poor　　　□ 59

20. In general how satisfied are you with your life?
　　　① Mostly Satisfied　　　② Partly Satisfied　　　③ Mostly Disappointed　　　④ Not Sure　　　□ 60

21. In general how strong are your social ties with your family and friends?
　　　① Very strong　　　② About Average　　　③ Weaker than average　　　④ Not sure　　　□ 61

22. How many hours of sleep do you usually get at night?
　　　① 6 hours or less　　　② 7 hours　　　③ 8 hours　　　④ 9 hours or more　　　□ 62

23. Have you suffered a serious personal loss or misfortune in the Past Year? (For example, a job loss, disability, divorce, separation, jail term, or the death of a close person)
　　　① Yes, one serious loss　　　② Yes, Two or More serious losses　　　③ No　　　□ 63

24. How often in the Past Year did you witness or become involved in a violent or potentially violent argument?
　　　① 4 or more times　　　② 2 or 3 times　　　③ Once or never　　　④ Not sure　　　□ 64

25. How many of the following things do you usually do?
- Hitch-hike or pick up hitch-hikers
- Carry a gun or knife for protection
- Keep a gun at home for protection
- Criticize or argue with strangers
- Live or work at night in a high-crime area
- Seek entertainment at night in high-crime areas or bars

　　　① 3 or more　　　② 1 or 2　　　③ None　　　④ Not sure　　　□ 65

26. Have you had a hysterectomy? (Women only)　　　① Yes　　　② No　　　③ Not sure　　　□ 66

27. How often do you have Pap Smear? (Women only)
　　　① At least once per year　　　② At least once every 3 years　　　③ More than 3 years apart
　　　④ Have never had one　　　⑤ Not sure　　　⑥ Not applicable　　　□ 67

28. Was your last Pap Smear Normal? (Women only)　　　① Yes　　　② No　　　③ Not sure　　　④ Not applicable　　　□ 68

29. Did your mother, sister or daughter have breast cancer? (Women only)　　　① Yes　　　② No　　　③ Not sure　　　□ 69

30. How often do you examine your breasts for lumps? (Women only)
　　　① Monthly　　　② Once every few months　　　③ Rarely or never　　　□ 70

31. Have you ever completed a computerized Health Risk Appraisal Questionnaire like this one?
　　　① Yes　　　② No　　　③ Not Sure　　　□ 71

32. Current Marital Status　　　① Single (Never married)　　　② Married　　　③ Separated
　　　④ Widowed　　　⑤ Divorced　　　⑥ Other　　　□ 72

33. Schooling completed (One choice only)　　　① Did Not graduate from high school　　　② High School
　　　③ Some College　　　④ College or Professional Degree　　　□ 73

34. Employment Status　　　① Employed　　　② Unemployed
　　　③ Homemaker, Volunteer, or Student　　　④ Retired, Other　　　□ 74

35. Type of occupation (SKIP IF NOT APPLICABLE)
　　　① Professional, Technical, Manager, Official or Proprietor　　　② Clerical or Sales
　　　③ Craftsman, Foreman or Operative　　　④ Service or Laborer　　　□ 75

36. County of Current Resident (SKIP IF NOT KNOWN)
　　　⑨⑨⑨ Other　　　□□□ 76-78

37. State of Current Residence　　　⑨⑨ Other　　　□□ 79-80)

Figure 2.5. Continued

These expectations will then become part of the actual plan for implementing the corporate wellness program. For example, one of the expectations may be to reduce the number of smokers by half. That then becomes a goal. The plan of action may include instituting a smoking policy, setting up smoking cessation classes, and monitoring the results.

The objectives and the plan can be divided into short- and long-term elements. A business leader must wonder about the risks of taking action and ask what the company has to lose or gain. If the company does nothing, health care costs will continue to consume a huge chunk of the profits.

Some action can be taken by the chief executive officer with little commitment of money or personnel time while the planning process is still in its early stages. The SANE approach, introduced in chapter 3, illustrates the ease and simplicity of getting started.

Once the corporate program is begun, more attention can be given to long-range planning (planning for periods of one year or more). Long-range planning will help solidify the general direction of the program, aid in constructing an annual budget, and provide a time line in which to schedule presenters, workshops, and facilities. Figure 2.6 shows a suggested long-range program schedule.

MANAGEMENT TEAM

Who will plan the program, implement it, and evaluate the process and outcomes? Much depends on the size of the company and, to some extent, on its monetary commitment. Once the corporate decision makers give the approval to begin looking into health promotion programming, the actual duties are usually delegated to a lower-level manager or to a committee of employees—the management team.

Responsibility for overseeing the wellness program could fall to one person who really wants to take charge of it. Ideally, the wellness work should be seen as part of that person's job, not assigned in addition to regular duties. If someone tried to take on responsibility for a wellness program and do their regular job at the same time, often the wellness work would take second place. It's important that management see the merits of assigning someone to handle wellness—with the same sense of importance as regular job duties. Or a committee can be appointed to make the decisions about wellness programs. A more expensive alternative would be for the company to bring in an outside consultant to handle wellness. In smaller companies, one or two people can perform the tasks. In larger companies, committees seem to work well. Active committees should be composed of wellness advocates—true believers—but not too many health nuts.

In some companies supervision of the program becomes a natural extension of the human resources, personnel, or training divisions. In

PROGRAMS	Jan.	Feb.	Mar.	Apr.	May	June	July	Aug.	Sept.	Oct.	Nov.	Dec.
Health Risk Appraisal	X	X	X	X	X	X	X	X	X	X	X	X
Aerobics	X	X	X	X	X	X	X	X	X	X	X	X
Brochure			X									
CPR		X		X					X		X	
Healthy Families Seminars	X	X	X								X	X
Hypertension	X	X	X	X	X	X	X	X	X	X	X	X
Leagues	X	X	X	X	X	X	X	X	X	X	X	X
Self-Development	X	X	X	X	X		X		X		X	
Smoking Cessation		X		X								
Stress Management		X	X						X	X	X	
Weight—Nutrition	X	X		X	X							X

Figure 2.6. Typical program schedule.

others, it is a logical role for the medical director or company nurse. Of course, some companies will want to hire a wellness director or even a wellness staff. But sufficient commitments of money to create positions usually occur—if at all—after the activity has grown to the point where the company feels the need for better coordination. Even if a company plans to hire a wellness director at any stage, employee involvement and ownership is vital to real success.

Find the True Believers

Employees involved in wellness programming should be true believers. True believers readily accept the wellness concept and the responsibility of designing, carrying out, and evaluating a worksite program. Besides being true believers, these staff members should also have specific qualifications to insure that the program will have a great chance of being highly successful. Employees with the least workload may not have the interest, ability, or the dedication to plan wellness programs. A good rule of thumb is that if employees are active and productive in their respective jobs, they will also be active and productive with health promotion efforts. Also, being good role models at the worksite will greatly enhance the committee members' effectiveness among their co-workers.

At Catholic Mutual Group, a small Midwestern insurance business, the CEO himself took on the job of investigating what to do about wellness for the 41 employees. The CEO—an ex-smoker—was joined in the task by another top manager and the personnel director.

True believers may either be joggers, employees who exercise regularly and stay fit by eating right, ex-smokers, recovering alcoholics, people who have recovered from a heart attack or are managing high blood pressure, or people who have a personal commitment to health and well-being. These interested people will work for the cause and sell it to others in the company. Compare the true believer with someone who may have been assigned to serve on the committee or who was delegated the task by top management. The levels of enthusiasm and commitment will be much different.

Some companies may choose to hire a health promotion professional to guide the program and chair the committee. This certainly is not a necessity, but companies with sophisticated, comprehensive programs may want to consider this alternative. Naturally, money is a big consideration in this decision. Of course, most companies are simply too small to consider hiring an outside person to be a full-time wellness programmer. The point to be made in this book is that a company can get into wellness without an elaborate committee structure and without spending much money at all.

The committee members should have a continuing and direct link to the corporate decision makers. This open channel of communication, along with the active support of the CEO and other top management personnel, is vital to worksite programming. Such support lends credibility and visibility to committee decisions. That doesn't mean the CEO should be out exercising aerobically with the troops. Some do. But even a memo of support from the CEO to all personnel goes a long way to say, "This company cares about you."

Where committees are used, a wide variety of selection processes and personnel are already in place. As an example of the committee form of decision making, the State of Nebraska has a large steering committee called the WellTeam. The WellTeam is composed of sixty volunteers representing nearly every state agency and the over 17,000 state employees. Members of the WellTeam not only act as a liaison to their own agency but serve as a pilot-testing group because they represent a cross-section of state employees.

WellTeam members also serve on subteams, which are smaller, working groups. The WellSaid subteam handles communication and advertising of the wellness programs. The WellPlanned subteam selects programs and providers based on surveys of employees' needs and interests. Because the employees are located throughout the state, the subteam contracts with community resources (like the YMCA's exercise program) that are available to most employees wherever they live and work. The WellDone subteam evaluates the ongoing programs and updates employee interest surveys.

Whether the committee's membership is three or 13, basic guiding principles need to be considered in the selection process. First, consider how many members make up a workable committee. In some cases three members may be too few, and in others 50 may be just right. The main point here is that the committee must have well-defined objectives and that the chairman be a good leader. This will help ensure the success of the committee.

Second, try to get a wide representation from the various departments. This becomes more important the larger the company is. Representation from key departments such as personnel is essential. Any medical staff, of course, should be involved. If the company is a union shop, it would be wise to have at least one representative from labor. The wellness committee at Valmont Industries, a medium-sized manufacturing company near Omaha, includes white- and blue-collar workers from shipping, customer service, finance, engineering, and sales.

Once the management team is established, it will have to deal with three vital issues before getting too involved in the planning process: (1) getting the commitment of top management; (2) deciding the best ways to keep employees informed; and (3) cultivating employee ownership.

Without management support, the wellness program is doomed. So it is vital to seek and obtain top management's commitment and involvement. At the very least, management will need to commit support services (personnel time, supplies, postage, computer time), facilities (use of conference rooms, empty offices, lunchrooms), and money (materials, programs, incentives). The management team can assume that top management will be willing to make this commitment since they approved the establishment of the management team in the first place. However, it is better to reconfirm that commitment than to risk having difficulty later on.

Another area to define early is whether top management support is active or passive. Active support means that the CEO or executive decision maker pledges support services, facilities, and money for the program—within some defined limits. Active top managers must also participate in corporate wellness activities, encourage employees to participate, and model a healthy lifestyle. Sometimes the most avid wellness supporters are the CEOs who are personally controlling high blood pressure, have stopped smoking, or have survived a near-fatal medical problem and are making lifestyle changes.

Passive support comes from a CEO who, for example, commits the resources without becoming personally involved. In fact, the CEO may be overweight, a heavy smoker, or a heart attack waiting to happen. But that same CEO can make wellness happen for the employees—even though he or she doesn't exhibit a healthy lifestyle. Management's personal involvement will help sell the concept and program to the employees.

Support from middle management and supervisors can be a continuing battle, even when top management (the CEO) is behind the wellness movement. Managers and supervisors often feel threatened when employees are allowed to use work time to participate in worksite programs. On the one hand, these managers feel pressed to get the work of the company completed, yet they think these programs disrupt the flow of work during a regular workday. On the other hand, they may want to go along with what they may perceive as a new venture by the company, and, at the same time, they want to let employees take part.

Once these managers realize that there is a payoff for them as managers (happier and healthier employees) and for the company as a whole (savings on health care costs), they may be more responsive and supportive.

Another issue is addressed by Blanchard and Tager in *Working Well*:

> One of the difficulties with introducing a wellness program to managers is that it often forces them to confront their own unhealthy habits. Depending upon how the program is introduced, this confrontation can ignite resistance, mainly because few of us want to be told that we are "wrong." On the one hand, we all know that living a healthy lifestyle is

important; on the other, most of us have at least some unhealthy habits. How can a health program succeed if the most influential members of an organization, its managers, smoke cigarettes, are overweight, and don't get enough exercise? The answer is that at the worksite, health is more than a personal matter, it is an organization objective and a management issue.

Open Channels of Communication

The second vital issue for the management team is determining what channels of communication to use to inform management and employees about the programs. Some of the traditional channels will be in place and should be used. These may include interoffice memos, newsletters, regularly scheduled staff meetings, bulletin boards, and word of mouth.

Other means of communication may be used to complement the traditional channels. Orientation meetings to inform employees of the start of the program or to introduce new material during the implementation stage has wide popularity. Usually all employees are invited to these informational meetings, which are held on company time.

Blue Cross/Blue Shield of Nebraska, for example, held an orientation meeting for all employees when the company introduced its plan to implement a smoking policy. The CEO, an ex-smoker, gave a testimonial and motivational talk, and the executive director of the Wellness Council told about the status of smoking policies in corporate America. The company's wellness committee (the management team) then outlined the yearlong schedule for phasing out cigarette machines and restricting smoking throughout the building. The final phase, a total ban on smoking in the building, is now in effect for the over 400 employees and anyone else, including clients.

Another channel of communication is the awards meeting, sometimes labeled the motivation meeting, annual meeting, or thank you meeting. This may also take the form of a picnic, a breakfast or other meal, or a party. The outcomes of the meetings are similar. The purpose is to introduce the wellness program or to discuss pending changes in programs already underway. Some companies use this meeting to reward people who have accomplished wellness goals (such as awarding prizes to people who quit smoking, to the department that collectively lost the most weight, or to the division of the company that had the most participants in a CPR class).

Western Printing in Omaha, for example, held an in-plant luncheon for its 37 employees to thank them for using their health insurance wisely. Because health insurance claims for the company were down in 1985, the insurance carrier gave Western Printing an attractive rebate on the premiums, which the company shared with the employees.

Other channels of communication may include specialty publications (payroll inserts on health topics, informational letters, subscriptions to national health magazines sent to employees at home), top management and management team visibility (the vice-president leading the company's team in the corporate run, matching T-shirts with the company logo for participants), and testimonials (naming of a wellness employee of the month, personal interviews with people who have made lifestyle changes published in the company newsletter).

Everyone Owns the Programs

The third vital issue for the management team is cultivating employee ownership. Unless employees buy into the idea, many wellness programs fail. A corporate wellness program may start on the initiative of a few true believers, but employees from every rung on the corporate ladder must feel that they own the program for it to be successful. Although the key to employee ownership is a simple concept, the actual process is difficult—getting and keeping the employees involved.

Involvement may mean getting employees to serve on the management team, creating subcommittees for employees in the areas of communication or evaluation, enlisting employees to teach programs such as CPR and first aid, or asking employees to do research for the committees or perform easy tasks. The more the employees are satisfactorily involved (to the extent they want to be involved in addition to their day-to-day work with the company), the more likely they will feel it is their program—not a take-it-or-leave-it program handed down from the executive suite. Highly successful programs boast of strong employee ownership from the beginning and at every step along the way.

TIME LINE

Create a time line to map out the long-range plan. Figure 2.6, introduced earlier, illustrates a yearly time line for a sophisticated program. In doing long-range planning, specific dates do not have to be identified months in advance. In fact it is best to identify only the month, and then work out a specific date as the time draws near. Such a strategy allows for the flexibility needed to schedule around other events such as work patterns, market downturns, production booms, product changes, management shifts, and other unforeseeable situations.

Notice that the summer months are less heavily scheduled because people are, presumably, spending time in outdoor activities. Some programs like aerobics continue throughout the year. Leagues are seasonal—softball in summer, bowling in winter.

FINANCIAL ANALYSIS

A corporate wellness program can be initiated for little or no money. Through the use of internal resources and outside nonprofit organizations, program costs can be kept to a minimum. On the other hand, as is the case with other goods and services, a substantial amount of money can be spent for personnel, equipment, and outside services from health promotion providers. The amount of money to be spent will be directly related to the goals and objectives the corporation wants to fulfill (identified in the strategic planning stage) and, naturally, the amount of money available.

Unlike other business purchases, what a company spends on wellness is not necessarily related to quality of programs or degree of acceptance by the employees. In other words, a company can spend a lot of money buying health promotion and never see a return on that investment. Employees may not be involved, and therefore, participation will be low. The principle of "you get what you pay for" does not hold true for health promotion. All the expensive exercise equipment in the world will not make employees healthier. In fact, some of the most successful health promotion programs (success being measured by number of participants, pounds lost, smokers who have quit, and health insurance claims reduced) are those that cost the least and use the expertise that is available within the company (an employee who is certified to teach CPR) and within the community (smoking cessation programs offered by the American Cancer Society).

Programs in WELCOM member companies are explained at the end of chapter 3 to illustrate the broad spectrum that exists in corporate America. Do not make the mistake of equating the inexpensive approach with small business and the more costly approach with large business. Keep in mind that the determining factor in financial expenditures is based on goals and availability of money, not necessarily on size of the company.

A worksite program can be started and maintained on a shoestring. The SANE approach introduced in the next chapter can go a long way on little money toward implementing a viable worksite program. In essence, money for health promotion programs, or the lack of it, is not an excuse for failing to begin a worksite program. On the contrary, a progressive CEO will realize that the benefits outweigh any costs involved.

A word of caution here: The issue is cost-benefit. What does a wellness program cost? And what are the benefits? There is a risk in conducting a program too cheaply if that means the program lacks quality. It may, in fact, not pay off and fail to live up to the high expectations. Don't overlook the importance of quality programming—regardless of cost.

Once the strategic plan is established, the management team needs to assign approximate dollar values to each phase of the plan. This can be accomplished by contacting vendors to get a rough estimate of the cost of those services and equipment in the market place.

Membership in a Wellness Council can make this stage easier. The member company can turn to the executive director of the Wellness Council to ask for recommendations of vendors, and find out what is available in the community. For example: Does our community have affiliates of the American Cancer Society, the American Lung Association, the American Red Cross? What other companies have used these programs, and how successful were they? Who is the contact person at these vendors? What other programs are available and what do they cost?

Forecasting

Vendors, once contacted, are often happy to provide estimates of costs. The management team might be surprised to find out just how many good programs are available at no cost or at very low cost.

When the items have been assigned dollar values and the total has been tallied, the management team must determine whether it is worth spending that money. In most cases the management team will have some idea of the amount of money that is available for health promotion programs and will know what yearly budget they can spend. A CEO might say, "See what you can do for $500." Within the constraints of that amount, a management team can look at the results of the surveys on employee interests and needs, compare those interests and needs with what is available from the vendors, and plug those programs into the budget. More often than not, the projected cost will be near the projected budget figure.

If the projected figure is over the amount of money available because the budgeted amount was not known or because of some last-minute programming additions, the management team has a few alternatives to consider. The first is to explore whether the program as planned can be done more inexpensively. This will require the management team to research internal as well as external options. For example, instead of having a for-profit provider present the smoking cessation classes, find a nonprofit organization that can accomplish the task for less money.

If the projected figure is still high, then the next approach is to justify the added expenditure by documenting for top management just how much money the company will save in reduced health insurance costs, reduced absenteeism and sick leave, and enhanced employee morale. It's hard to put dollar figures on things like morale, but every company knows just how much it costs each day an employee is sick. And companies are finding out that employees who smoke cost the company from $300 to $600 or more, even up to $4,600 more per year, than employees who don't smoke.

Don't overlook the idea of asking employees to share the cost of the programs with the company. Free programs are sometimes taken for granted. Some health promotion experts say that splitting costs provides an even stronger incentive for the employee to participate ("I'm spending my own money."). Northwestern Bell initially asks employees to pay for a smoking cessation program. If the employee does not smoke at work for six months, the company reimburses the employee for the entire cost of the program.

If top management sets aside a certain amount and says, "See what you can do with $500," then the management team must work with that figure, review the plan, set priorities, and revise.

Cost Versus Savings

To find out how much money is saved by instituting wellness programs, a company needs to know how much sickness and lost work days have been costing the company over the last few years. Figure 2.7 provides a simple formula for figuring out what the company pays for sickness. Have the personnel director fill in the numbers.

The other part of the financial analysis is to retain accurate records concerning the expenditures. These records will be used along with the information compiled in the internal market analysis to help determine the cost savings to the corporation. Other statistics such as the health care costs of smokers as compared with nonsmokers or the costs involved in recruiting and training new personnel may also be helpful.

Trying to determine the corporate cost savings of a worksite program is a difficult task. It is especially hard to factor out the effects of redesigning the benefits package (short-term savings are obtained by increasing deductibles and shifting costs from the company to the employees). Changes in the health care delivery system in the market place also make it difficult to determine just how much money a company is saving. For example, if employees join HMOs, actual costs of care are not available because employees pay a fixed amount per year for health care.

AUDIT

Evaluation is too often the forgotten part of a corporate wellness program.

Evaluation does not have to be extensive, expensive, or conducted by experts. Whether the program is simple or comprehensive, an evaluation procedure can be implemented to fit the resources and needs of the

Sick Care Costs Audit

1. HEALTH INSURANCE PREMIUMS

19____   19____   19____   19____   19____

(Five years ago)   (Current year)

_____

Single
Coverage
_____

Family
Coverage
_____

Costs of Not Taking Action

Percent change between years one and five = _____.

Assuming a constant rate of change, by 19____, (five years from now), we can expect to pay $_____ for the same number of employees—an increase of $_____ in the next five years.

2. DISABILITY PAYMENTS

Workers' Compensation

Current year
Total $_____

Average per
employee $_____

Social Security Disability

Current year
Total $_____

Total per
employee $_____

3. ABSENTEEISM

Total Days of
Absenteeism (19____, _____
current year)
times (X) Average
cost per day   _____   =   _____

PLUS (+) Total cost of absenteeism   _____

Total health insurance   _____

Total disability   _____

Miscellaneous costs   _____

_____

TOTAL SICK CARE COSTS   = $ _____

Figure 2.7. Sick care costs audit. Adapted and used with permission from *Wellness at the School Worksite,* Health Insurance Association of America, Washington, D.C.

corporation. Something as simple as counting heads, dropouts, pounds lost, or miles jogged takes no special talent or use of machines or formulas. Corporate experience has been that new worksite programs start with simple evaluation procedures but gain in complexity as they expand.

Both process (Did we do what we said we'd do, on time?) and outcome (Did we improve employee health and reduce costs of health care?) evaluations are essential for a number of reasons. First, they provide the participants with reports of their progress and, consequently, provide a source of personal incentive. Second, evaluations provide the presenter (the management team and the outside vendor, if one is used) with a means of charting the progress and effectiveness of the programs. Third, evaluation gives the corporation a chance to examine the presentations, materials, and presenters. Fourth, good evaluation may provide top management with a guideline for financial support. And finally, the results can be entered into the already established vast body of knowledge to reinforce the positive results of worksite health promotion everywhere.

Evaluation procedures can be as simple as asking employees what they think about the wellness program. In a confidential written questionnaire at Central States Health & Life, 84 percent of the over 400 employees reported that they felt worksite wellness was "good for employee morale." This is especially significant because at the time not quite 60 percent had participated in any way in any wellness program. The conclusion that management drew was that even if, for whatever reason, employees chose not to participate in worksite programs, they felt good about a company that made such programs available.

The more complex evaluation procedures include physical fitness measurements, blood chemistries, body fat composition analyses, blood pressure screening, and cost analyses. These procedures cost money, use personnel time, and usually require sophisticated equipment to perform. In some cases, the tests can be performed in-house by corporate personnel with the use of corporate equipment. Obviously, large companies have medical staff on board who can do these tests as part of their regular work schedule. Smaller companies do not have medical staff or equipment. This should not discourage corporations and smaller companies that do not have these resources.

Community resources like the local universities, hospitals, YMCAs, or Visiting Nurses Association may assist or may provide all the personnel and equipment needed to take the measurements. The costs of these services may be deferred or reduced through grants, volunteer help (students, interns), monetary gifts, or agreements with organizations to exchange the privilege to test corporate employees for use of the results in research and professional publications.

Knowing the difficulty of determining if wellness programs are working, many companies simply decide not to spend money evaluating but to put that money into more programs and let the good results just happen. Of course, careful monitoring of employee needs and interests (this means circulating another round of employee surveys to measure their needs and interests) will go a long way to finding out exactly what employees want and need. Also, careful monitoring of the sick care costs audit, by plugging in new figures each year for what a company pays for health care insurance claims and sick days lost, may reveal some strong short-term gains on the bottom line.

Researchers tell us that the only real way to evaluate whether worksite health promotion is working is to do the costly and complex scientific evaluations. These evaluations would take place over a long period of time, and employees would be measured in a "before and after" assessment. Small businesses aren't equipped, or willing, to undergo such scrutiny, nor do they need to. But the larger corporations are able to spend the money to evaluate programs, and their medical staffs are set up to tally the numbers and measure the body fat. The rest of corporate America then can reap the benefits of their findings, often through membership in Wellness Councils and by serving on committees in the Wellness Councils.

The point to be made in this book and with this business plan is that a company, large or small, does not have to spend a lot of money getting into wellness or a lot of money trying to figure out if wellness is paying off. By following the simple procedures outlined in this business plan, any business can do what it does best: Make a company and industry analysis, perform an internal market analysis, do some strategic planning, appoint a management team, project a time line, set aside a budget, and audit the results.

3

THE SANE APPROACH TO A HEALTHY COMPANY: SMOKING, ALCOHOL, NUTRITION, EXERCISE

The SANE approach gives the CEO four specific things to do right away—with no commitment of money and little added personnel time—but sets the stage for more detailed, more expensive health promotion programs.

Let's discuss each of the categories in the program and show, in detail, what the CEO can do to promote wellness in these areas. From the SANE framework spring other programs. Figure 3.1 shows the branching of the programs from the no-cost, low-cost suggestions involving smoking, alcohol, nutrition, and exercise.

WRITE A SMOKING POLICY

Any CEO or decision maker can sit down at his or her desk today and write a smoking policy for the company. Whether it's a five-person office, a manufacturing plant, an electrical utility, or a retail store, a smoking policy can be written that will respect the rights of the smoker and the nonsmoker.

The memo that went to employees of Central States Health & Life Companies is shown in figure 3.2. It is based on the "Model Policy for Smoking in the Workplace," a sensible booklet that answers questions about how to go about respecting everyone's rights. The booklet, which

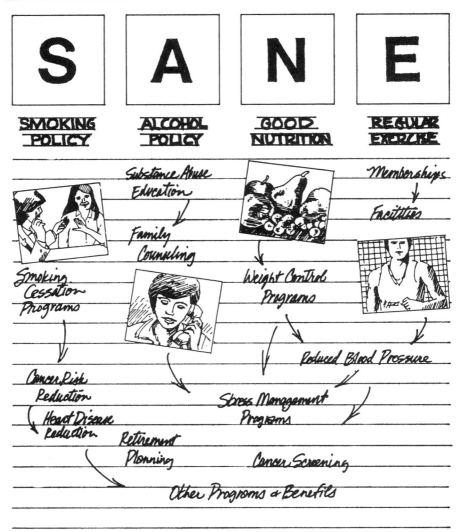

Figure 3.1. The SANE approach to a healthy company.

includes suggestions for appropriate wording, is available directly from your local American Cancer Society unit.

Also, the American Lung Association publishes two helpful booklets for business: "Taking Executive Action" makes the case for recognizing nonsmoking as the norm; and "Creating Your Company Policy" provides a handy employee survey form and guidelines for writing a policy. The Association offers several versions of smoking policies and discusses the problems of ventilation, air purification, smoking in shared work spaces, and ongoing enforcement and conflict resolution.

Some members of the Wellness Council regard enforcing a smoking policy as an impossible task. Television news reporters, for example, are prime examples of high pressure employees on a deadline—and smoking is almost as mandatory as knowing how to type. Other difficult areas in which to tackle the smoking issue would be for union workers, union shops, and employees who work away from a central office or on construction sites.

With employees working in the field and in company vehicles, writing and enforcing smoking guidelines gets tricky. Yet the Metropolitan Utilities District—Omaha's gas and water utility—successfully phased in a number of smoking guidelines over a period of one year. For field staff who work together in company-owned vehicles, if all occupants of the vehicles are smokers, smoking is permitted. If one employee is a smoker and the other is not, the vehicle is a nonsmoking area. In construction shacks and trailers, smoking and nonsmoking areas are to be designated.

On the other extreme of the smoking issue, WELCOM member Dana Larson Roubal & Associates, an architectural firm with about 100 employees, has a no-smoking policy in effect. Although the written policy states that employees can request that certain areas be designated as "smoking permitted," the decision makers have not approved any such areas, thus creating a smoke-free building.

Smoking cessation programs help employees stop smoking. Companies use a variety of reimbursement plans. Some pay the entire cost if the employees stay cigarette-free for, say, one year. Others split the cost of the program with the employee who pays through payroll deductions.

Sometimes companies offer smoking cessation programs before a new smoking policy goes into effect. At the same time, companies try to have the cigarette vending machines removed from the lobby or break rooms.

Here are many other low-cost or no-cost ideas your company may want to try. Companies that are members of the Wellness Council of the Midlands (WELCOM) have contributed the following ideas, and they are credited here:

> Participate in the Great American Smokeout by contacting your local unit of the American Cancer Society. The smokeout is usually held in November of each year. Individual companies can provide "smoking survival stations" equipped with sunflower seeds, carrot sticks, and sugarless gum for participants. The Smokeout is designed to create a smoke-free environment for one day. Follow-up programs may be offered to help employees quit for good. Also,

Central States Health & Life Co. of Omaha

TO: All Personnel

RE: Change in Policy Regarding Smoking

As a convenience to smokers, Central States has always allowed smoking
throughout its quarters. The Company also recognizes a self-serving
purpose by allowing smoking at one's work station in that valuable time
might be conserved.

In recent years, however, reliable medical data have been accumulated
revealing previously unsubstantiated concerns regarding the risks to
nonsmokers subjected to a smoke environment. Measureable levels of
nicotine have been found in the blood and urine of nonsmokers exposed to
tobacco smoke. Such exposure presents a special health hazard.
Breathing the smoke of others can lead to unsafe carbon monoxide levels,
allergic reactions, and exacerbation of conditions such as asthma and
bronchitis.

As a result, the management can no longer fail to recognize the rights
of nonsmokers. There are valued members within our work force with
serious respiratory ailments and allergies. For these people, cigarette
smoking is not just an annoyance, but an additional hazard to their
health and well-being.

In its statement of purpose, the Company has committed itself to (1)
provide an atmosphere of efficient, cheerful employment; and (2)
encourage at all times a sense of fair play and justice in all things.
It is in the spirit of our avowed purpose that we appeal to those among
us who smoke to willingly accept our change in policy.

Accordingly, effective (date), smoking will not be permitted in working
areas of the Company. Smoking will be permitted in designated
nonworking areas, which will be equipped with exhaust devices to carry
smoke out of the building. In this manner we believe we are providing
equal rights to smokers and nonsmokers alike.

For many, these changes, instituted for the common good, will be
frustrating and very difficult. To the nonsmokers we ask for continued
tolerance and forebearance for their associates most affected by this
change. Lastly, we appeal to everyone working at CSO to join in helping
to bring about this change in as friendly a manner as possible.

Should anyone have a question, please feel free to visit with your
supervisor or (contact person's name) in Personnel.

96th and Western • Omaha, Nebraska 68114 • Phone: (402) 397-1111
Mailing Address: P.O. Box 34350 • Omaha, Nebraska 68134-0350
A CENTRAL STATES OF OMAHA COMPANY

Figure 3.2. Memorandum outlining a smoking policy.

offer an exercise alternative during lunch or before lunch ("Exercise Your Butts Off," Central States of Omaha).

Serve breakfast to smokers who are participating in the Smokeout. Have them leave their cigarette packs on the table when they leave (Blue Cross/Blue Shield of Nebraska).

For smokers who have stayed cigarette-free for six days after the Smokeout (and only 3 percent actually quit for more than that one day), give them a turkey for Thanksgiving (Blue Cross/Blue Shield of Nebraska).

Adopt smokers and be their moral support throughout the day of the Smokeout (Central States of Omaha).

Offer cold turkey sandwiches to smokers who pledge to quit during the Smokeout.

Make a contribution to the American Lung Association as a tribute to employees who have stopped smoking (Millard Manufacturing).

Frame a smoker's last pack of cigarettes and send flowers each year to commemorate the anniversary someone quit smoking (Alco Container Corporation of Omaha).

Conduct your own smokeout day or designate one day (Friday) as no-smoking day each week.

Make a commitment to keep ashtrays clean as a first step in a smoking awareness campaign (Omaha Federation of Labor).

In educational environments, such as elementary and secondary schools, allow smoking only in areas well away from students because it is crucial for teachers to exhibit positive lifestyle examples (Creighton Preparatory School).

Reward employees who have stopped smoking. (Valmont Industries held a recognition dinner called "I Kicked the Habit." Those who had quit for over a year were eligible to win a three-day vacation and to receive 32 hours of pay.)

Install smoke-removing machines which remove by-products of smoking. These are best placed in smoking areas of cafeterias, conference rooms, and hallways where smoking is allowed.

When designing new facilities, talk with architects and mechanical engineers about design features and air-handling equipment that minimize problems with smoke from cigarettes.

Remove ashtrays from conference rooms, offices, reception areas, restrooms, and public entrances. Or put wrapped candy in the ashtrays.

Put up posters extolling the virtues of not smoking. Many are available free from voluntary health agencies or the federal government. (The chancellor at the University of Nebraska Medical Center made up his own. Under a picture of the chancellor himself pointing a finger in Uncle Sam fashion, Chancellor Charles Andrews is quoted as saying, "I want YOU to cut it out!")

TAKE A STAND ABOUT ALCOHOL USE AT COMPANY EVENTS

Write a policy regarding the use and serving of alcohol at company-sponsored events. Employees get mixed messages when they are provided an employee assistance program to help them deal with alcohol and drug use, yet the company president announces that "the bar is open" after a particularly harried business meeting. What about those wine and cheese receptions that seem so elegant in late afternoon but offer no alternative drinks such as juice for the employees who don't want to or can't drink alcoholic beverages?

Some guidelines for use of alcohol at company-sponsored social functions appear in figure 3.3.

Alcohol and drug misuse often endangers not only the abuser but other people as well. In working situations in which others can be put in jeopardy, some companies address those issues in written policies. The Omaha Public Power District, for example, asks supervisors and fellow employees to be constantly alert for employees who may be using and abusing drugs or alcohol—especially when the safety of other employees, the public, or the company's equipment or reputation is at stake.

Drinking and driving can be deadly companions. A company's alcohol policy should also be backed up with a statement on use of seat belts. An example of a simple policy is given in figure 3.4.

Here are more no-cost or low-cost ideas for your company when it comes to alcohol and drug awareness.

Invite Alcoholics Anonymous to use your meeting rooms for their sessions.

Invite a local alcohol counselor for a brown-bag lunch session with your employees.

At wine and cheese receptions, serve nonalcoholic beverages as a visible and acceptable alternative, or don't serve alcoholic beverages at all.

Devise a nonalcoholic but tasty holiday punch for the next office party.

Distribute nonalcoholic party drink recipes.

Conduct seminars for parents of teenagers on teen drinking and drug use.

Have a local expert speak on wise use of prescription and nonprescription drugs and interactions of medications.

Encourage employees to produce and act in a psychodrama about alcohol and its devastating effects on an employee's job and family (Central States Health & Life Company of Omaha).

Policy Administration Manual

Guidelines for Company-Sponsored Social Functions

Purpose: To Provide Guidelines to Aid in the Planning for, and the Supervision of, Behavior at Company Events Where the Consumption of Alcohol is a Possibility.

Our Company now recognizes the role all employees play by setting an example regarding the use of alcohol at Company functions. To fulfill this role in an effective manner, we are committed to emphasize the moderate use, as well as non-use, of alcohol at Company-sponsored events. This commitment is founded on the need for consideration for those who do not use alcohol; consideration for those whose lives are affected by the misuse of alcohol; and consideration for those who encounter difficulty in their personal use of alcohol. No employee, or family member, should ever feel pressured or coerced to consume alcohol in any form at Company functions.

The following guidelines should be used when planning and/or supervising Company-sponsored events where alcohol may be consumed:

1. Alcohol is not essential to any Company-sponsored recreation or business activity.

2. The use of alcohol is a personal choice. No one should feel pressured to drink, or not to drink, or be made to feel uneasy or embarrassed as a result of their choice.

3. The person responsible for organizing any event where alcoholic beverages are to be served is accountable for seeing that alcohol is served in a manner which respects its potential as a drug which affects safety and health.

4. When alcohol will be used at Company-sponsored events, the following circumstances apply:

   a. Non-alcoholic beverages should always be visibly available in adequate variety and supply and should be presented in as appealing a manner as beverages containing alcohol.

   b. Food, such as dry snacks and hors d'oeuvres, should always be provided with alcoholic beverages.

Figure 3.3. Guidelines for company-sponsored social functions (alcohol use). Adapted from *Responsible Hosting Guidelines* © 1985. Used with permission from Northwestern Bell Telephone Company.

Policy Administration Manual

Guidelines for Company-Sponsored Social Functions (cont'd.)

   c. Bartenders should be instructed to use moderate to light amounts of alcohol in mixing drinks. Arrangements where the liquor provider is paid by the empty bottle are to be avoided.

   d. "Cocktail" or "happy hours" or "attitude adjustment" periods should generally be scheduled for 45 minutes or less. The bar should close promptly at that time.

   e. Permit each person to accept or decline an alcoholic beverage. Avoid assumptions of what people want to drink at meals.

   f. Where wine is served, it should be only a complement to the meal, and not an event marked by repeated servings.

   g. The Company manager hosting or responsible for the function will ensure that these guidelines are observed.

   h. When people have to drive automobiles home following a Company-sponsored function, no after-dinner alcoholic beverages should be offered or served.

5. If the hosting manager is aware that a guest has had too much to drink, he/she should attempt to provide supervised transportation home for that person. (This guideline also holds true for a person hosting a company-related function in the home. The hosting person is liable for their guests, particularly in case of an accident.)

6. When representatives of the Company are present at functions sponsored by another company where alcohol is available, professional and responsible behavior and the use of good judgment is expected.

Figure 3.3. Continued

74

Policy Administration Manual

Seat Belt Usage

**Purpose: To Promote Employee Safety
When Traveling in a Company Car.**

Health and accident research shows that fastening one's seat belt is the single most important preventive safety measure that Americans can take to reduce disability and save lives.

For this reason, the Company requires anyone traveling in a company car to wear a seat belt. This applies even to short distances.

Because we are concerned for our employees' safety and well-being, the Company encourages regular use of seat belts at all times.

Figure 3.4. Policy on seat belt usage.

COUNT CALORIES AND MAKE NUTRITIOUS CHOICES

Many American workers grab a few minutes during the day for lunch and usually grab a sandwich or chips at the same time. Then, while answering the phone, reading the newspaper, or working on the computer, the busy worker will balance a soft drink and french fries on a desk piled high with paperwork. It might be time to take a look at the luncheon menu and the American way of eating lunch at work.

If the company has a cafeteria (and most small or medium-sized businesses don't) or if your employees dine in a common eating facility in a building you share with other companies, ask the food supplier to post the calories of the meals, have the menus checked by a qualified dietitian, and suggest putting in a salad bar for build-your-own concoctions. Sometimes helping nutrition in your office might be as simple as buying a small refrigerator to put in the file room for lunches your employees bring from home.

If you have vending machines in your company or in your office building, discuss with the vendor the possibility of offering alternative selections of healthful foods. For example, alongside the Twinkies and Cheetos place sugar-free gum and sunflower seeds.

Under the category of nutrition and good eating habits are several suggestions for companies of any size:

Put a picnic table outside.

Invite Weight Watchers to use your meeting rooms for their sessions. The Weight Watchers at Work program is tailored for working people. It addresses specific topics such as coffee breaks, surviving lunch hour, fitting exercise into a busy schedule, and the effects of time pressure and stress on eating habits (Blue Cross/ Blue Shield of Nebraska). (Creighton University, through the School of Nursing, offers on-campus Weight Watchers programs for all employees and students.)

Print a better health shopping list. The plastic-coated card should list healthful food items; tell the shopper which foods contain complex and simple carbohydrates, protein, and fat; and should stick to the refrigerator with a magnet. A shopper can check needed items in pencil and erase the list for reuse (Northwestern Bell).

Buy a microwave oven for employees to use. It increases the kinds of foods they can bring for lunch.

Put a scale by the coffee pot or in the restrooms.

Provide salt and sugar substitutes in your lunchroom.

Offer decaffeinated coffee too.

Start a vending-machine awareness program. Post the sodium and caloric contents of the items in the vending machines (and the safe levels for daily consumption). Write directly to the companies that manufacture the products for accurate content information. (The State of Nebraska employees began a program called "You're in Control." Colorful table tents list the contents of the items in the vending machines, and the tents are placed on tables in each canteen area of the state office complex. The lists, which are changed monthly, are good ways to publicize the wellness program.)

Announce a "healthy box lunch" day and have a nutritious meal brought in. Invite a local expert to sit in and discuss good eating habits (Alco Container Corporation).

Instead of serving sugary carmel rolls at your next business meeting, substitute bran muffins and fresh fruit. Try popcorn and fruit juices for refreshing breaks during meetings.

LET FITNESS THRIVE THROUGH EXERCISE

Exercise is the fourth component of the SANE approach, and it is the stepping stone to more sophisticated wellness programs. As a CEO, by getting the word out that you will allow organized exercise to take place, say, in the conference room, you are giving your employees the message that "it's OK." Some companies already have organized exercise and don't realize it. All those volleyball and softball teams are evidence of that. But employees are cautioned to get into shape to participate on those teams.

For business leaders who still think they have to provide showers for employees, exercising could be conducted after work. Those who want to bring in a video player and work out to a Jane Fonda tape or invite an aerobics-minded employee to lead the routines simply jump into their sweats in the restrooms at the end of the workday and go home for showers.

The management attitude that allows exercise to begin leads to other organized activities. Soon a group interested in jogging together might meet after work, or walkers might take noon jaunts around a nearby park in nice weather and around an enclosed shopping mall in bad weather. All these activities begin to create a healthy company.

Here are some more exercise ideas that won't cost much.

Suggest that employees walk to work (maybe only one day a week). Park at the far end of the parking lot.

Push equipment aside in the shop and hold exercise sessions after work (Lozier Corporation).

Sponsor a fun run for employees and their families. Supply T-shirts as prizes.

Turn an unused office into a quiet room where employees can take a quiet break and relax.

Hold exercise classes in the lunchroom after work. Find out if any of the employees are certified Jazzercise instructors (Blue Cross/Blue Shield of Nebraska).

Buy aerobic exercise cassettes or videotapes for employees to use at home or at work.

Suggest that employees use the stairs instead of the elevators.

Take stretch breaks at your desk during the morning and afternoon.

Buy or supplement memberships in nearby health clubs for employees or reimburse them for going if they can show a record of attendance.

Provide maps of walking and jogging areas near your workplace. Indicate the flat, semi-hilly, and hilly routes and time frames to complete them within the lunch hour (Alco Container Corporation).

Noontime walking programs can be organized. Determine scenic routes, watch for hazards such as broken sidewalks and barking dogs, distribute informational sheets on stretching exercises to do before walking and on how to calculate a resting and target heart rate. Provide pulsemeters for walkers to measure their heart rates before and after walking. (Steppin' Out is the noontime walking program for employees of the State of Nebraska. Early participants received certificates for being charter members. A group of walkers who prefer to stroll for 30 minutes at lunch has formed as well.)

Arrange corporate discounts for employees and their families at stores that sell exercise equipment, warm-up clothes, and athletic shoes.

Encourage employees to ride bikes to work. Install bike racks.

Form a group called "Exercisers Anonymous" and take daily walks around a downtown mall or indoor shopping center (employees in the law firm of Kutak, Rock & Campbell).

Support employees who join volleyball, baseball, and softball leagues (Lutheran Medical Center).

Employees who walk or run 100 miles, swim 50 miles, skate 250 miles, or bike 500 miles receive membership in the company's Century Club (membership is open to over 900 employees of the Lozier Corporation, a manufacturer of store fixtures).

If you have an exercise facility and your company is open twenty-four hours a day, try to keep the facility open all day and all night (Saint Joseph Hospital).

Once a company has created a healthy environment, other activities naturally follow. The SANE approach encourages companies to do what they can with the resources they have. The following ideas are taken from WELCOM's member companies. Use them in good health:

Designate a bulletin board for health information only (Health Connect at Millard Manufacturing).

Send away for informational brochures from government agencies such as the National Health Information Clearinghouse (P.O. Box 1133, Washington, DC 20013-1133, phone (800) 336-4797). Many of the publications are free or can be reprinted without copyright restrictions.

Subscribe to health magazines and newsletters and circulate them in the office or plant or have them available in the break room.

Subscribe to health magazines and have them sent to employees at home (the Metropolitan Utilities District pays for subscriptions for 770 employees).

Build a library of self-help books and cookbooks. Let employees borrow them.

Form support groups of employees composed of women who are pregnant, people who are trying to stop smoking, those who plan to lose weight, or those who are interested in exercising.

Buy a blood pressure machine and train someone to use it. Encourage employees to have their blood pressure taken often. People who are monitoring hypertension will be pleased to have such a service available.

Buy relaxation tapes and self-help cassettes. Let employees check out video and cassette players overnight and over weekends to use the tapes.

Have the company logo or name put on T-shirts, towels, and socks. Use these as incentives or rewards.

Look to your employees and their families for expertise. Someone may already be trained by the American Red Cross to teach CPR or first aid. Consider having someone trained on company time and at company expense. That person can then teach the other employees. Post names of certified employees; they may be called on in an emergency.

Ask the personnel director or the person who handles health insurance claims to give a workshop on how to be a better health care consumer.

Hold brown-bag lunches once a month and invite local experts to speak on the topics of nutrition and weight control for busy people, willpower against smoking, alcohol abuse, first aid techniques, and balancing career and family (Wellness on a Shoestring program, Blue Cross/Blue Shield of Nebraska).

Work with local newspapers to write a feature story on your innovative new wellness program.

Put together a mobile exhibit to travel around to different plant sites or work locations. The exhibit can display informational materials about the company's wellness programs and incentives such as T-shirts and towels or gift certificates.

Run a contest for employees to submit ideas for a company wellness logo. Use the logo on T-shirts, towels, headbands, socks, shoelaces, and other incentive items. (The State of Nebraska proposes to put its logo on stickers to be displayed on state cars.)

Offer noontime seminars on women's health issues including breast self-exams, working mothers, and working while pregnant. Other seminars could cover general-interest topics such as when to call a doctor, medical self-help, first aid, CPR, and seat belt use. The time slot between 3 and 4 p.m. may be used for seminars too. If the company has flex hours, midafternoon may not conflict with work, car pools, or child care.

Conduct health risk appraisals and medical screenings for glaucoma, high blood pressure, diabetes, sickle cell anemia.

Have your company support a health fair and encourage employees to attend.

PROGRAMS OR FACILITIES? A DETAILED LOOK AT BOTH

The employees of healthy companies often take the ball and run with it after the corporate culture has evolved to support them in making their own lifestyle changes. Each company can develop its programs in a number of ways. Sometimes a company will be program oriented and bring in educational and informational speakers, seminars, materials, and workshops. Other companies may become more facilities oriented and look into purchasing exercise equipment or having employees build their own. Ideally, a combination of programs and facilities (even if those facilities are health clubs your employees belong to) works best to promote wellness in any worksite.

Whether your company becomes program oriented or facilities oriented or some combination of the two, it is useful to look at examples of other companies for ideas. The following section is a discussion of a sophisticated incentive program at Central States Health & Life Company of Omaha that rewards exercise and also activities that are not fitness oriented, like wearing a seat belt. Also discussed is Mutual of Omaha, which offers a range of programs in-house and contracts with nearby health clubs to provide workout facilities.

Millard Manufacturing sets an example as a company that is not facilities oriented but has a strong wellness program underway and is seeing bottom-line results when it comes to fewer lost workdays. Labor and management working together to reduce workers' compensation claims became the impetus for innovative programs at Valmont Industries and Midlands Hospital. Both companies showed significant dollar gains, and their wellness programs are discussed in some detail.

Model examples of facilities-oriented programs are found at HNG/InterNorth now Enron, an energy company with several hundred employees, not counting their families, in Omaha. Contrast that sophisticated facility with the modest exercise room at Kutak, Rock & Campbell, a law firm with about 200 associates and support staff.

One other type of sophisticated wellness program needs to be discussed, and that is the community-based arrangement. Under this program, a company contracts with a local provider, like the YMCA or a health club, to do fitness testing and to provide facilities in the community for employees and their families to use. IBM is an example of a major corporation whose Plan for Life uses community resources in cities throughout the United States, wherever IBM employees work. The alliance between Union Pacific Railroad and the Downtown YMCA in Omaha is a prime example of how big business can support a community program and vice versa. Also discussed is the community link between Creighton University and the Omaha Police Department.

Go for the Gold at Central States

In an effort to reward employees who seek to improve their health habits, incentives are a big part of wellness at Central States Health & Life Company of Omaha (CSO). In fact, CSO's incentive programs—Go for the Gold and Health for Wealth—are among the most innovative and novel programs in any company. Designed as a six-month incentive program, CSO's Go for the Gold ran in conjunction with the 1984 Winter and Summer Olympics. Employees earned medals to cash in for prizes. The ultimate prize was a gold medal paperweight and $75.

The rewards weren't easy to achieve. An employee had to earn ten bronze medals to receive a silver medal, and five silver medals were needed to get the gold. Regular exercise and attendance at informational seminars were the backbone of the program. Forty-five employees won gold medals. That's over 10 percent of a 350-plus work force. One was a 38-year-old vice-president who used to smoke three packs of cigarettes and drink as many as thirty cups of coffee a day. He credits the "enthusiasm and positive atmosphere created by the wellness program" as being his primary reinforcement to quit smoking for good and to reduce his coffee consumption.

Another employee, a stressed out working mother, rearranged her day to fit aerobic exercise in at 5:30 a.m. She also works out at the company's noontime classes. She lost weight, got control of her diet, and won a gold medal for her efforts. Even a vice-president in a branch office in California won a gold medal. He established a no-smoking policy in the branch office and keeps himself conditioned at age 48 by lifting weights and doing aerobics. But perhaps his most important commitment to the wellness of others is the donation of blood every two months.

Because the program included so many areas in which to win medals, employees of all fitness levels and at all stages of wellness awareness could win. Such an equitable program builds in ways of rewarding everyone, not just the secretary who jogs at lunch or the supply clerk who lifts weights, but the executive who quits smoking and the programmer who wears a seat belt while driving.

Another CSO incentive program that lasted one year was called Health for Wealth. Again, employees were rewarded for practicing healthy habits and could earn $10 a month in addition to sportswear imprinted with the company logo. Employees who participated used a special log on which to record their monthly accomplishments in the areas of aerobic and nonaerobic activity, seminars attended, blood donations, weight, nonsmoking record, and seat belt use.

HealthWorks at Mutual of Omaha

Because Mutual of Omaha provides plans to meet higher health care costs for its clients, Mutual has a real incentive to promote wellness and to educate employees about good health.

With the support of V. J. Skutt, the CEO and one of the original founders of WELCOM, a 15-member wellness committee developed the following mission statement to direct their efforts at the world's largest individual and family health insurance company: "The mission of the wellness committee is guided by the conviction that employees who pursue a lifestyle of wellness can live longer, healthier, and more productive lives. As a result, the Companies can realize improved worksite productivity, lowered absenteeism, and reduced health care costs."

To emphasize the committee's efforts, a distinctive HealthWorks symbol has been adopted to provide easy recognition of all wellness issues. For example, a special HealthWorks menu has been added to both dining areas in the Home Office. The menu includes 20 different meals in a four-week cycle. Each meal contains about 500 calories. The HealthWorks menu also complements the Nutrition Connection, a 16-session weight loss program held twice a week during lunch periods. Classes are taught by a registered dietitian.

Many wellness activities are coordinated through the Home Office's health service facility. Available at no cost to employees and staffed by registered nurses, Health Service provides treatment for minor illnesses, administers first aid, and coordinates blood pressure screening programs and employee participation in a smoking cessation clinic.

For employees who need guidance in pulling their personal wellness programs together, the company subsidizes participation in an Omaha hospital's Physical Fitness/Positive Living course. Classes meet twice a week for twelve weeks and address topics such as exercise, nutrition, stress management, habit control, and motivation.

Because space at the Home Office is limited, the company has not established an exercise facility for employees; however, arrangements have been made with nearby health clubs that allow employees to keep fit at discount rates.

While promoting wellness inside the Home Office, the Mutual organization remains active in citywide and regional wellness activities. As an annual host to the Health Fair of the Midlands, the Home Office becomes a forum for health education and screenings. The company also sponsors a 10K road race and two-mile fun run during the Health Fair. The race has already established itself as one of Omaha's premier running events.

Muffins and Juice at Millard Manufacturing

Manufacturing plants present their own set of health and safety hazards, but the owners of Millard Manufacturing made great strides in protecting the safety of their 45 employees and in educating them about wellness. The company fabricates metal food-processing equipment for industry. Ron and Sue Parks, representing a new breed of American small-business owners, made major modifications in the shop environment in the areas of temperature control, ventilation, and lighting. The claims for workers' compensation took a sharp nosedive at the same time. They recognize that a well-trained and motivated work force is the key to success in business.

Along with safety, the Parkses brought wellness into the shop and office, and they are seeing results. Many employees have quit smoking; smoking is banned in the office and production areas anyway. But they continue to keep health education materials in front of their employees in the company newsletter and on a bulletin board that is designated specifically for health and wellness information.

Exercise facilities are not in the plans for Millard Manufacturing, but this company is seeing results from its educational campaign. One incentive program in particular has contributed to employee productivity: The company rewards employees when they have no unexcused absences or tardies for one week with muffins and juice. It seems like a small reward (a healthful one, however, because the company began the program using coffee and donuts), but muffins and juice work at Millard Manufacturing. And employees are not absent or late much anymore when compared to their past track record.

Wellness Pays Off at Valmont

If Millard Manufacturing represents a typical small business, then Valmont Industries is an example of a medium-sized American company. The company employs 2,000 people worldwide, but the manufacturing plant and headquarters are located in the small town of Valley, Nebr., just west of Omaha. Valmont manufactures mechanized irrigation equipment, steel poles for area lighting and electrical distribution, and light wall pipe and tubing. Valmont also provides a series of financial services for its customer base and has a personal computer distribution business.

The 1980s have been tough for farmers—the prime buyers of Valmont's mechanized irrigation equipment—and the market downturn in agriculture has taken its toll on Valmont as well. So why does a company put money into wellness programs at the same time it is laying off workers?

The uncertainty of the market for their products and wondering who was going to have a job from week to week was putting a lot of stress on the company's valuable employees. Coupled with what the company called an overly generous health insurance plan, skyrocketing costs of health insurance, and a rising accident rate resulting in increased medical costs and payments of workers' compensation claims, the company decided to do something to bring down costs.

Management and labor instituted a safety awareness program called Operation Pride which focuses on each individual's personal responsibility for maintaining health and working safety. Employee education is central to the company's efforts to control costs and improve the quality of life for the employees and for Valmont as a business.

The net result of these efforts by labor and management has been that Valmont's medical costs showed a 3 percent decrease in 1983. In 1984 the company continued that downward trend with costs down an additional 2 percent per capita. Turnover rates among plant personnel and exempt and nonexempt administrative staff went from 9.7 percent in 1981 to 5.4 percent in 1984. Another interesting figure is the company's workers' compensation costs. In 1981, claims totaled $256,000; in 1984, by comparison, workers' compensation claims were $161,000.

So when anyone asks if wellness is paying off at Valmont, the vice-president of human resources, Tom Whalen, cites the numbers to prove it.

No More Aching Backs at Midlands Hospital

Ninety percent of the workers' compensation claims at Midlands Community Hospital were for back injuries. Certainly hospital staff have to lift many things, including patients and heavy equipment. Back injuries seemed inevitable. But when it paid more than $100,000 in claims for back injuries in one year, the 500-employee hospital took definitive action to cut its losses.

The hospital instituted a two-tiered program: (1) preemployment back screenings of high-risk applicants for jobs requiring physical activity, and (2) a back clinic for assessment and education of current employees who had ever filed a workers' compensation claim for back injury. Applicants for hospital positions that might require them to lift, carry, and bend are assessed in the physical therapy department. If the physical therapist determines that the applicant is not suited for the job physically, and if the job cannot reasonably be modified, the applicant is not hired for that job. Employees who have had back problems in the past are required to attend a two-hour back clinic in the physical therapy department. There they are taught the proper body mechanics and monitored to correct bad habits in lifting.

The program netted an average annual saving of $50,000, and the dramatic decline in back claims and repeat injuries produced not only substantial bottom-line payoff but healthier, happier hospital employees.

Join the Enron Private Fitness Center

When Enron, an energy corporation, purchased some old office buildings in downtown Omaha, one of those buildings contained a swimming pool and gymnasium, an indoor track, racquetball courts, and locker rooms. The company renovated the facility and turned it over to the employees, who set up a private fitness center for employees and their families.

Of Enron's over 2,000 eligible employees, 27 percent have elected to belong by paying modest dues each month. Payment of dues entitles employees and their family members to participate in the wellness activities and to use the facilities. Professional staff provide instruction in a number of areas including fitness evaluations, first aid, diet, weight training, aerobic dance, and overall wellness.

Even Lawyers Exercise

When Kutak, Rock & Campbell, a law firm, renovated its offices in a historic building in downtown Omaha, the partners decided to put in an exercise facility complete with a racquetball court. The program has evolved to include a portable basket over the racquetball court which enabled staff to participate in basketball games. Other staff use the court for aerobic workouts and wallyball.

Fit-Check

When the Union Pacific Railroad wanted to offer a program that would help their employees find out about fitness, they did what millions of Americans do each year—they called the YMCA. In fact, a unique national program was devised by JoAnn Eickhoff at the Downtown YMCA in Omaha in response to a request from the railroad's assistant vice-president of personnel, Vern Krider. What resulted was a program called Fit-Check.

Fit-Check is a comprehensive health enhancement program administered by the staff at the YMCA. Employees of Union Pacific, who want to participate and have approval from a physician, fill out a health questionnaire, are given a fitness test battery by trained professionals in physical education at the Y, attend an orientation session at which they receive results and an exercise prescription, and are given a retest to chart their progress.

Because Union Pacific employs hundreds of people throughout most of the western half of the United States, it was important for the company to set up programs for employees who worked in other sites. By working with a community resource like the YMCA, which has similar operations in cities throughout the country, Union Pacific employees could be offered the same program as the people at corporate offices in Omaha, no matter where they live and work. So employees in Los Angeles, Salt Lake City, St. Louis, and eight other cities were offered the same program—coordinated by the Omaha YMCA and administered in YMCAs in those other cities.

The partnership has fared well. Union Pacific claims to have healthier, more productive people with less absenteeism. Overall, the participants in the retest had higher fitness levels in areas such as cardiovascular endurance, muscular strength, flexibility, heart rate, and blood pressure. And the YMCA is marketing a unique program that can work for other companies that want to take part in Fit-Check.

Fit Police

The City of Omaha, through its Health Monitoring and Physical Fitness

Program, noted an increase in productivity and a one-fourth decrease in sick leave used by 100 police officers who took part in an assessment developed by the physical education and exercise science department at Creighton University. Creighton is working with the Omaha Police Department to evaluate cardiac risk factors, aerobic capacity, and exercise training of these city employees who, by the nature of their work, are at high risk for stress and its related illnesses.

Six months after the personalized exercise programs began, the officers who participated in the pilot test were better able to handle physical work and prolonged stress. This included a drop in coronary risk factors. The city plans to put the entire police force through the program.

MOVING FROM THE COMPANY TO THE COMMUNITY

Have you joined the ranks of the true believers? Are you convinced that wellness works? Business executives make high-level decisions every day. Sometimes they make those decisions based on less evidence and proof than we presently have for the merits of worksite wellness. But the numbers are starting to speak for themselves. Employees are living more healthful lifestyles, and employers are benefiting at the same time.

This book now makes the transition from the company to the community. Wellness at work spills over into the community in more ways than one. Workers who learn about nutrition and exercise at work take those ideas home with them. Families everywhere benefit in a number of direct ways from programs conducted at worksites across the country.

Businesses are joining together in cities of every size to share their experiences and to learn from each other. Companies are coming together to form Wellness Councils. And that's what's so new about this book. Part II tells decision makers just how Wellness Councils, through the efforts of business and industry, create a healthy community.

PART TWO

4

WELLNESS COUNCILS CREATE A HEALTHY COMMUNITY

The mission of Wellness Councils is to promote wellness at the worksite. This mission is guided by the convictions that the quality of life in any community is in large degree measured by the health of its people, and that as more employers offer quality programs, not only will more individuals benefit, but the spiraling costs of health care can be better contained.

ROLES OF WELLNESS COUNCILS

Councils promote worksite programs and bring employers together with community health care providers and other resources already in existence. In that sense, there is a good symbiotic relationship between the councils and the community service providers (such as the American Cancer Society, YMCAs, and other nonprofit health organizations) and vendors (such as health clubs, hospitals, and stores that sell exercise equipment). Ideally, the professional staff of the Wellness Council helps an employer devise an appropriate program—determine what is needed, like a stress management program—and then the council links the employer with a local provider. In this example, it might be a hospital program in stress management or a mental health association.

Councils contribute to creating a healthy community by playing specific roles for corporations and business leaders, small business owners, workers and their families, and the community in general. Here's a look at what councils do for each of these consumer groups.

Corporations and Business Leaders

For its corporate members the Wellness Council acts as a clearing-house of information about other successful programs, including documentation and analysis of program effectiveness. The Wellness Council assists employers in selecting programs that both fit their needs and are within their budgets.

For companies that want to set up worksite wellness programs and just don't know where to begin, the Wellness Council provides tested guidelines so that the companies—large and especially small—can assess their needs and make informed decisions about what programs their employees want and what programs the company is able to afford. That's what the business plan presented in chapter 2 is all about. In fact, it's almost like having the executive director of the Wellness Council take your company through the decision-making steps toward designing a wellness program.

For companies that have programs in place, the business plan offers ways to enhance or expand those programs, suggests ways to involve more employees or set up mechanisms to reward success through cash and other incentives, and helps companies evaluate what community-based programs will suit their needs. The business plan presented in chapter 2 also tells how to evaluate a program to see if it is making a difference—in health care costs, in reduced smoking, in increased awareness, in pounds lost by employees, in reduced absenteeism, in calories counted, in laps jogged.

This is the sort of information that grows out of the collaborative efforts undertaken by companies that join Wellness Councils. Providing local expertise and knowing what works for other companies in the community and nationally; sharing ideas; publishing guides, newsletters, and payroll inserts; holding delegate meetings and annual meetings; bringing in national experts on wellness topics—that's what the Wellness Council does for its corporate members.

Small Business

In some ways small businesses have more to gain from participation in Wellness Councils than do large businesses. There is no doubt that larger companies have more resources to spend. Many times the large companies are willing to take a risk and experiment; sometimes they will even spend money to evaluate a program's effectiveness. By working on the same committees, attending the same meetings, and sitting around the luncheon table with representatives from big businesses, small business people can benefit from the experiences of the large companies. So the pooled information shared in the Wellness Council benefits all businesses, but especially small businesses.

The owner of a small business (a printing company, for instance) can serve on the same committee with the personnel director of a large utility who directs a megabuck wellness program. In effect, the large company takes the risks, makes the investments, evaluates, and makes the results available to every member of the Wellness Council. Sometimes the results are shared as part of formal programs. Many times, however, they are shared through the informal networking that takes place at Wellness Council functions.

Wellness Councils may also bring small businesses together to share the cost of worksite wellness programs that they can't afford by themselves. For example, a six-person business with two smokers, in cooperation with five other companies, may be able to collect enough people to attend a smoking cessation program, and even get a price break from a vendor.

Across the country, large office complexes are being built in major urban centers. The hundreds of small companies that lease space in those properties could easily join together to plan wellness programs and share costs. That involvement could range from a stress management class held in one of the larger conference rooms to construction of a facility for exercise and perhaps even child care.

As the wellness movement continues to grow, small business entrepreneurs who are true believers in wellness will, no doubt, discover innovative ways to make money on wellness.

Workers and Their Families

The Wellness Council was founded on the belief that the key to living a healthy life depends on knowledge and good information and a nurturing, supportive environment. Not only do Wellness Councils provide the needed information and knowledge, but they put employers in touch with health-service agencies that are eager to share their information through materials, audio-visual presentations, and workshops designed specifically for a particular company's work force.

Speakers at delegate meetings sometimes address specific family issues such as aerobic exercise for parents to do with children, parenting skills, and family self-esteem. In WELCOM's Corporate Wellness Series of fitness events, family members and children participate in equal partnership with employees from member companies.

The Community

In Omaha alone, the Wellness Council touches the lives of over 70,000 people who work in businesses throughout the Midlands. Add to that their family members, and the total number of people affected by the

Wellness Council is about one-third of the people living in the combined metropolitan areas of Omaha and Lincoln, Nebraska, and Council Bluffs, Iowa (estimated to be nearly a million people).

Through business and industry, any community has its best opportunity to create a well-managed and exciting wellness environment. In such an environment, the Wellness Council has become a powerful voice, even the conscience of the community, in matters of health promotion. The Wellness Council continues to promote existing community health resources and lobbies for the creation of additional resources (jogging and bike trails, for example) which contribute to the physical, mental, social, emotional, occupational, and spiritual well-being of everyone in the community.

The Wellness Capital of the World

The Wellness Council continues to reach out into the community to create a healthy atmosphere. Here are some activities the Omaha council endorses:

> The Health Fair of the Midlands (an annual health event sponsored by a local TV station, a major insurance company, an energy company, and the American Red Cross)
>
> The airing of 30-second public service announcements covering topics like the risks of smoking and high blood pressure
>
> Open invitations to the public to attend various health promotion seminars
>
> The sponsorship of the Corporate Wellness Series
>
> Co-sponsorship of a national health promotion conference

In fact, WELCOM's written goal is to foster a total community environment that, through its commitment to worksite health promotion, will result in the Midlands being recognized as the Wellness Capital of the World.

WHAT WELLNESS COUNCILS ARE AND WHAT THEY ARE NOT

Wellness Councils have been confused with many things. But the best way to describe them is to liken them to chambers of commerce in which businesses in all sectors of the economy—no matter what product or service they sell, without regard to how many or how few people they employ—work together for a common goal. In fact, Wellness Councils and chambers of commerce are natural allies and can cooperate with each other in many areas.

In promoting wellness programs at the workplace, a Wellness Council acts as a catalyst to get worksite wellness programs started in as many companies as possible within the community. One of the most important things a Wellness Council does as a catalyst is to gain the support of the chief executive officers in the business community. Second, the Wellness Council acts as a counselor to individual businesses who want to initiate programs or build on programs that already exist. One of the council's primary functions is to act as a clearinghouse of information.

Full details about the day-to-day operations of a Wellness Council and ideas for Wellness Council activities (such as corporate runs and conferences) appear in the next chapter.

Wellness Councils Do Not Compete

Wellness Councils do not provide on-site wellness programs for companies. They are not in competition with the nonprofit groups such as the YMCA or the for-profit vendors such as franchised wellness packages. In fact, Wellness Councils enhance the business climate for such groups. Here's how that works: A member company may ask the Wellness Council about setting up a program on smoking cessation. A staff member or volunteers from the Wellness Council generally meet with the company seeking assistance to try to determine what the company wants to accomplish, how much it wants to spend, and how long it wants to make a commitment to smoking cessation.

Then the council provides the member with a list of vendors whose programs are approved by the Wellness Council's medical advisory committee and lets the company decide which vendor to work with. The Wellness Council will then follow up by giving the member company a list of other companies who have had experience with those recommended smoking cessation programs.

Wellness Councils are not involved with employee assistance programs (EAP), which are designed to provide confidential help for employees who have problems, especially those related to drug and alcohol abuse. Businesses have discovered that it is more cost-effective to help their employees cope with personal problems than it is to go through the never-ending cycle of hiring, training, and termination—or worse yet, allowing an impaired employee to affect sometimes critical day-to-day operations. Wellness programs are the preventive side of employee assistance programs and do not take the place of EAPs. Rather, wellness programs give employees information and the opportunity to make life-changing decisions that ultimately may prevent many problems from getting to the EAP. As is the case with most social problems, prevention of drug and alcohol abuse is always more cost-efficient than treatment.

Similarly, Wellness Councils are not safety councils. Local chapters of the National Safety Council are natural allies of any Wellness Council, however. Safety councils provide information to employers about good safety practices at work. Although there is some overlap in their missions, the focus of the safety council is to prevent accidents at work. Wellness Councils strive to create an environment that makes it easier for people to make decisions and adopt behaviors that lead to healthier lifestyles and, consequently, a more meaningful existence.

Occupational safety codes—regulated by the federal government—are also a part of a healthy workplace but not a specific concern of Wellness Councils.

Cost Containment—The Quick Fix

In Nader-like fashion, groups of local businesses in 105 communities (according to the U.S. Chamber of Commerce) have banded together to form cost-containment coalitions with the sole purpose of reducing health care costs.

These cost-containment coalitions collect data and determine what the pricing schedules are for doctors and hospitals. They measure length of hospital stays, for example, or frequency of doctor visits. Once they have compiled these figures, the coalitions share the data with coalition members and try to negotiate better rates through group discounts. Sometimes they create enough public pressure to bring down health care costs. The mission of these watchdogs is a short-term, quick-fix approach to reducing health care costs. But coalitions may get reduced rates only for their members. Unless costs are actually reduced, the costs are merely shifted to companies that aren't members of the coalition.

As a result of cost-containment coalitions throughout the country, the public has a new awareness. People are slowly starting to shop for health care as carefully as they do for a new car. Hospitals are competing for patients, and even doctors are beginning to advertise. The recent surge of activity in HMOs and PPOs may, in some part, be a result of the cost-containment movement.

Perhaps one of the greatest measures to come out of the cost-containment movement is the second opinion on surgery. Every business person knows that in any field there are some unscrupulous people who will abuse their position. In medicine there are honest differences in treatment as well. If surgery is indicated for someone in your family, the value of a second opinion—measured in emotional distress and dollars saved—is worth it.

Wellness Councils, rather than being a short-term quick fix, are the long-term answer to cost containment. In fact, many cost-containment coalitions, once they have had some degree of success, have found the promotion of worksite wellness to be the logical next step. Indeed, some Wellness Councils have grown out of the organizational structure of cost-containment coalitions.

For WELCOM, members of the local cost-containment coalitions and their natural adversaries like hospital administrators and insurance executives have found themselves working side by side to promote worksite wellness through the Wellness Council.

Chapter 5 gives a detailed look at how Wellness Councils work, from founding to formation and growth.

5

WELLNESS COUNCILS: HOW THEY WORK

What you are doing in the Wellness Council responds to this Administration's belief that private initiatives rather than government programs can be of great benefit to the nation in dealing with many social and community needs.

—President Ronald Reagan
Personal letter to the author
December 1981

The Wellness Council of the Midlands (WELCOM) is the pioneer of business initiatives in which companies join together to promote good health in the workplace. In no other community group—except possibly the local Chamber of Commerce—do business associates and competitors in the marketplace sit side by side and conduct business from which everyone benefits.

There was no blueprint for success when WELCOM began. So its founders, in their wisdom, imposed few rules and regulations on the newly formed group, choosing instead to set up easily revised bylaws. The full text of the bylaws and articles of incorporation is reproduced in appendix A. An official organization chart appears later in this chapter.

No matter how a Wellness Council is formed, the day-to-day operations can be generally patterned after the WELCOM model. Some councils, like WELCOM, begin because a CEO is willing to take the lead. The Health Insurance Association of America is especially interested in starting councils in this manner and has appointed a coordinator to recruit CEOs of large insurance companies in cities across the country.

The councils in Norristown, Penn., and Greensboro, N.C., began because executives in insurance companies took on the responsibility to get other CEOs involved.

Wellness Councils, as discussed earlier, can grow out of the work of cost-containment coalitions. And the councils in Minneapolis, Columbus, Ohio; Atlanta; and Chattanooga, Tenn., are examples. A third way for councils to begin is for a local business leader who is not from the insurance industry to take the lead. Such a council is well underway in Tucson, Ariz., because of the efforts of a private health spa owner. Wellness Councils might also spring from the local Chamber of Commerce or safety council, but these routes are as yet untested.

THE FOUNDING STAGE

In a 1980 report, the government set goals for the nation's health and urged employers to take the lead in establishing local initiatives:

> Health officials and health providers must be joined by employers, labor unions, community leaders, school teachers, communications executives, architects and engineers, and many others in efforts to prevent disease and promote health. It is important to emphasize that, while the Federal Government must bear responsibility for leading, catalyzing and providing strategic support for these activities, the effort must be collective and it must have local roots. (Department of Health and Human Services, *Promoting Health/Preventing Disease: Objectives for the Nation,* 1980)

The goals set forth in *Promoting Health/Preventing Disease* continue to guide health promotion efforts around the country. Specific goals for employers to shoot for in the workplace are discussed in chapter 3. The discussion targets the areas of smoking, alcohol and drug use, nutrition, exercise, and stress management.

With that charge in mind, the formation committee of local business leaders, with the full support of the three CEOs of Omaha's biggest companies, formally invited the CEOs of other prominent, local companies to the kickoff luncheon in January 1982. It was an elite group. CEOs who couldn't come were not allowed to send a representative. Nor were general managers from national corporations invited (although they were later invited to join and did so with much enthusiasm). The idea was to get the support directly from local business owners and chief executives, not from their delegates or assistants.

The letter of invitation said, simply, please join us to hear a distinguished guest speaker and to "share our enthusiasm regarding formation of the Wellness Council." And the letter was signed by the big three business leaders who had already decided to support the Wellness Council and whose prestige and names lent high visibility to the movement.

From the opening remarks and the presentation by the guest speaker to the final appeal for support, the luncheon hosts should set the stage for the day's mission: to start a Wellness Council. A suggested agenda would include opening remarks by the CEO hosts, then lunch followed by the guest speaker and the final appeal for support from the business community. Business leaders need to know what the council is and how it will help their businesses.

Here are the opening remarks from the 1982 WELCOM luncheon:

> The Wellness Council is the product of a small but dedicated group of individuals, most of whom are active in the business community. After a great deal of research and study, this group of business people strongly believes that good health does add up to good business.
>
> The Wellness Council, therefore, believes that any efforts to improve the health and general wellness of the American work force is a wise investment, an investment that begins immediately to pay dividends and, most important, an investment, which although very tough to document, will nevertheless pay big dividends for years to come.
>
> Any informed discussion of health and wellness will emphasize two pillars of the wellness doctrine: (1) Wellness—good health—is primarily a personal choice and therefore is 100 percent the responsibility of the individual; and (2) a community working together can create an atmosphere that helps individuals help themselves.
>
> As leaders in the business community, our business interpretation of these two wellness principles is that because wellness is really a matter of taking personal responsibility for one's own health, choices related to life must always be voluntary. The Wellness Council believes that the private sector and not government is the logical provider of appropriate information, skills, and motivation. And because a supportive atmosphere greatly increases the chances for success, the Wellness Council believes that the workplace is by far the most logical setting to identify problems, transmit information, and to conduct action programs that will encourage those in the work force to accept greater responsibility for their own health.
>
> As business people we understand the bottom line. We will ask directly and tell you exactly how we need your support, but we know that any commitments—we like the word investments—any investments you make must be compared to actual benefits that result from your participation. The Wellness Council readily accepts that kind of accountability.

The formation committee invited Dr. Charles Berry to be the guest speaker because he had long been speaking out about the role of prevention and wellness. Dr. Berry had directed NASA's medical program when the effects of space on humans was still as uncharted as space itself. Dr. Berry, a Nobel-Prize nominee, went on to become founder and president of the National Foundation for the Prevention of Disease. The insurance industry often calls upon his medical expertise, hiring him as a consultant, and he serves on the medical advisory council to the insurance trade group.

Dr. Berry has participated as the keynote speaker in several Wellness Council kickoff luncheons, and his message is a powerful one. Here are some excerpts:

> In 1958, I was involved with selecting the original seven astronauts and from then on with getting man into space and back safely. Man's capability to work in space was unknown. Prestigious groups such as the National Academy of Sciences, congressional committees, and even the President's scientific advisory committee were concerned with whether man could survive even a fifteen-minute flight into space, let alone work there.

> They said we didn't have all the data, and certainly we did not; however, one will never have all the data. I strongly believed we might never have any more data if we did not act prudently on what we had and add to this data as we progressed. We went ahead.

> Today, the same "not enough data" charges are leveled at the pursuit of wellness. I want you to leave here today convinced that we do not know enough about the major risks to our health and the means to reduce these risks, and therefore dedicate yourselves to achieving wellness for yourself and your employees, and making this a part of your corporate or small business structure and life.

> In talking with corporate chief executive officers, boards of directors, or small business owners, my task is to supply the facts so that they may come to believe, as I do, that wellness programs at the worksite make sense, are good business, and provide benefits to both employees and employer.

> Some of the things we do know, which allow us to go forward now, relate to today's three top killers: heart disease, cancer, and accidents. They are not caused by bacteria and viruses as the killers at the beginning of the century were, but by agents we call risk factors. Those factors identified by extensive studies of heart disease are having a strong family history of heart attacks, total cholesterol levels above 200 mg/dl, high blood pressure, cigarette smoking, obesity, sugar intolerance, lack of exercise, stress, and use of oral contraceptives. Note, that all these but the family history are controllable, if you desire and work at it. Also by reducing the controllable risk factors, one can hold off the heredity. The really good news is that reducing your risk factors will stop progression of the deposits of fatty material in your arteries and even reverse the process. What an incentive! Interestingly, some of the same dietary habits that

prevent heart disease are also active in preventing cancers of the breast and colon.

Thus, it's time to make your own vaccine against today's killers. Identify your risk factors, then fill your syringe with the necessary amounts of reduced fat intake, weight control, increased fiber intake, smoking cessation, blood pressure reduction, exercise, and so forth.

Motivation is our biggest problem and will continue to be, for we must all be initially motivated and then repeatedly remotivated.

Risk reduction programs or intervention programs are also called health promotion programs and come in many varieties and packages. All companies are different and I believe these differences are best determined by going through steps to reach implementation of the intervention portion of the wellness program. These steps begin with CEO and management commitment, followed by a health audit of the company, and progressing through motivation of employees, health screening, and feedback of risk factor information confidentially. Next is combining audit and screening results in order to select the intervention programs and implement them.

The major impetus for looking at wellness programs today has been the skyrocketing cost of medical care and a company's inability to control this skyrocket. Contributing to this rise are the obvious insurance costs and the not-so-obvious but probably more important costs of absenteeism, coverage for ill employees, interruption of service, replacement, and reduced morale and productivity.

Healthy, happy and productive employees are a company's greatest asset on their balance sheet. Some CEOs recognize this and have said, "Whether it saves money or not, I believe it's the right thing to do for our people." Others have said, "We are cutting back so I can't start a new program now." In fact, you can't afford NOT to start it, for the benefit of your employees, their morale and for the future of your business.

While there is difficulty collecting the data to document the dollars saved by these programs, we are progressing in this task monthly. Aside from individual company savings already reported, I believe we will see a $2 return for every $1 invested in your employees through company wellness programs.

A well-versed guest speaker, such as Dr. Berry, can talk about health promotion and wellness in the language of business to business leaders. A strong speaker is the key to moving the business community to action.

Where Do We Start?

Many businesses, including many of those represented at the kickoff luncheon, may already be running wellness programs but perhaps are interested in conducting a greater variety of programs. Many other businesses are interested but do not know where to start. Finally, the Wellness Council believes that many companies will want to begin programs when they really understand all of the implications and potential benefits. The Wellness Council wants to help those interested companies begin or expand their programs.

The Wellness Council is a nonprofit, private corporation governed by a representative board of directors. It operates solely on private dollars and hires and supports a professional staff.

Funds for the council come entirely from private sources. Membership fees provide a major source of funds. The fee structure includes corporate memberships for companies both large and small. There is also a separate fee structure for nonprofits and educational institutions.

As the council grows and its programs solidify, its founders expect to be able to generate funds through program and workshop fees. Within a few years it is the council's goal to be entirely self-sufficient through funds generated by its membership fees and in program and workshop revenues.

In addition to membership promotion, board development, and fund raising, the primary role of the council staff will be to provide consultation services and technical assistance to member organizations as they set up, expand, and develop their in-house wellness programs. This may take the form of on-site, in-house assistance or special workshop or training experiences for council members.

The council will be very diligent not to duplicate or compete with the programs of existing groups, agencies, or companies. A great deal of important work is already being done in the community. The purpose of the Wellness Council is to enhance and help coordinate those efforts.

Creating an Epidemic of Wellness

In closing the luncheon, tell prospective members that they can catch the contagious excitement and show their support by sending a check and naming a delegate.

And as they walk out of the dining room, hand them a packet of information that contains a slick folder imprinted with the logo of the Wellness Council and the statement of purpose. Attendees at WELCOM's kickoff received a professionally designed and printed folder containing information about the formation committee members and a short history of the movement in Omaha to this point. The logo was adapted from Leonardo Da Vinci's *Immortal Man*.

But most important, the folder contained a registration card that served as an invitation to these CEOs to join the Wellness Council.

There was no pressure, no hard sell. In fact, if they didn't want to participate, they could show that by not doing anything.

The chief executive officers who were invited but did not or could not attend were not left out of the business of the day. Right after the luncheon, at great effort and expense, WELCOM delivered by courier service the same packet of information the attendees received.

Who Will Make the Commitment?

A corporate executive or other person who is thinking about assuming a leadership role must first ask these questions of himself or herself:

1. Do I have a personal conviction that wellness programs at the worksite are worthwhile and serve a valuable purpose?
2. Will I initiate a worksite wellness program in my own company, if one is not now in place?
3. Am I willing to provide energy and leadership to a community initiative of this kind through my own personal involvement?
4. Can I find the time to lead the movement in my community?
5. Would I be willing to stay with the project for as long as three years?
6. Can I gain the cooperation of my company to provide initial up-front resources such as
   a. Money for luncheons, wine and cheese parties, public gatherings, advances for honorariums and transportation
   b. Secretarial and mailing services
   c. Accounting services—collections, accounts payable, records
   d. Public relations and media activities
   e. Printed materials, stationery, and brochures
   f. Meeting rooms, parking
7. Am I confident that I can convince other important civic and political leaders to join with me as co-founders?
8. Do I have subordinates available
   a. To stay on top of activities, write speeches, take charge, handle telephone calls, draft agendas, and call and conduct meetings?
   b. With skills in health promotion to advise me, to be seen as our in-house experts, and who are willing to accept community involvement?

J. Kenneth Higdon, who directs the health insurance industry's Wellness Council project, travels around the country recruiting CEOs in major insurance companies to take the lead in Wellness Councils in their communities. Higdon, himself a veteran insurance executive and former president of Business Men's Assurance Company of Kansas City, maps out a cost-value equation for his colleagues.

On the value side, Higdon states, are three reasons why a CEO would want to put money and effort behind wellness at work and in the community: One, it is important for the CEO of a health insurance company to be perceived as an activist in the health care cost arena. Just as gas station owners bore the brunt of Americans' frustrations with rising oil prices in years past, the insurance companies are taking the flak for raising premiums in response to rising health care costs. By being in the forefront and addressing health care issues head on, the insurance company will be the good guy.

Two, wellness is a noncontroversial issue. If an insurance company executive were to take on doctors or hospitals in a head-to-head battle over high costs, that would be a risky business. But promoting wellness is an absolutely acceptable approach—as American as mom, the American flag, hot dogs, and apple pie (although hot dogs which are high in sodium nitrite and apple pie which is high in sugar and cholesterol may not be so great after all). Wellness, embraced by all segments of society, is a win-win proposition.

The third value hits the company right at the Home Office. A lot of people in the community are probably insured by the company for medical risks, so the company, along with making a positive financial impact on its own business, will enhance its position with the hometown crowd.

Everything has a price, and the astute CEO needs to know that such an undertaking will cost the CEO personally and will cost the company. Higdon outlines the three specific costs he puts on the cost side of the cost-value equation:

1. A commitment of personal time to seek out other business leaders to co-host a luncheon for other CEOs and major employers in the area

2. Payment for the luncheon and support staff to organize it and prepare the materials

3. Naming of a staff member within the company—someone who has a personal interest in wellness—to be what Higdon calls the "obstetrician" of the local Wellness Council. This key staffer must clear time during a suggested gestation period of nine months and must have full CEO support and access to support services within the company such as printing and secretarial services. Similar staff assignments are made, for example, for United Way campaigns. Sometimes the obstetrician is from the market-

ing department or public affairs, or is the medical director, personnel manager, or cost-containment adviser.

PRIVATE DOLLARS FOR MEMBERSHIP FEES

Although the Wellness Council is a nonprofit volunteer organization, its operational monies do not come from charitable contributions. Member organizations pay yearly membership fees based on employee population. The fee structure is $2 per employee with a minimum of $250 and a maximum of $2,000. Nonprofit organizations and educational institutions pay $250 per year regardless of how many employees they have.

To help with programmatic activities, WELCOM also received a grant from the Kiewit Foundation, a local philanthropic organization. These grant monies have helped WELCOM to publish its first newsletter, design and publish payroll inserts, produce public service announcements (PSAs) for television, and cover the costs of the annual meeting. Grant money has also helped defray the cost of staging the national wellness conference in Omaha and the council's Corporate Wellness Series.

Other funds are generated from the sale of some of WELCOM's publications and PSAs, although many of those items are priced at cost-recovery figures, and their purpose is not to generate revenue directly.

How to Become a Member

Word of mouth and selective solicitation provide WELCOM with new members. Each method provides about half of the new members. WELCOM's reputation and its commitment to worksite wellness help to market the organization to the business community. But the Wellness Council also uses the assistance of a retired executive to approach selected businesses.

The impetus generated at the kickoff luncheon continues to build, and local companies generally know what the Wellness Council is. Some view their membership as a valuable tool and an adjunct to their own ongoing wellness programs. Others see their membership as good public relations, much as they would see memberships in the local arts council, symphony, or ballet. Still other companies are simply curious and see a small investment of $250 as money well spent to explore the best thinking of other business leaders.

No matter what reason companies join, and regardless of the level of their commitment, every company benefits from the clearinghouse of information available through the Wellness Council. And on-site programs are begun and enhanced because the council's staff works in tandem with the delegates and CEOs.

The executive director decides which businesses to solicit for membership. The leads are provided by members, business journals, and newspaper stories. The retired executive makes the initial contact and works with the business until the decision maker is ready to set up a meeting with the executive director. Follow-up is a key step here.

Who Joins Wellness Councils?

WELCOM's membership reflects a cross-section of business and industry in the Midlands. All main segments of American business are represented: agribusiness, manufacturing, transportation, retail, utilities, insurance, service companies, and wholesale trade. Member companies range in size from a five-person medical office to the over 6,000-member work force at Mutual of Omaha's World Headquarters, to the 17,000 people who work for the State of Nebraska in agencies located throughout the state. Also represented are supermarkets, advertising agencies, retail stores, savings and loans, accountants, law firms, banks, construction companies, manufacturers, and utilities.

Some members fall into categories known as "providers" or "vendors." Providers include nonprofits such as YMCAs, the health service organizations such as the American Lung Association, and other voluntary health agencies such as the Visiting Nurses Association. Vendors are profit-making businesses such as Smokenders, hospitals, and doctors in private practice. Providers join for two reasons: One, they want to do business with other member companies, and two, they are themselves employers. They have employees for whom they design their own brand of wellness programs.

A spokesman for the Nebraska Division of the American Cancer Society says that through its membership in WELCOM, the Society is "better able to conduct its free adult education programs at the worksite." By targeting specific cancer-related deaths through worksite programs, the Society helps private industry reduce its losses. "We are able to reach a broad spectrum of the community, many of whom would otherwise miss our life-saving messages" (personal letter to the author).

Educational organizations such as universities, school districts, and private schools are active members. And government groups, such as the State of Nebraska and the City of Omaha, represent thousands of government workers and their families on the council. Figure 5.1 graphically shows the areas of cooperation.

What follows is an informal account of just how the Wellness Council of the Midlands is structured and how it works. Other Wellness

CONSUMER, RELIGIOUS, CIVIC AND SOCIAL ORGANIZATIONS

BUSINESS AND INDUSTRY

VOLUNTARY HEALTH AGENCIES

Wellness Council

EDUCATIONAL ORGANIZATIONS

PROVIDERS OF HEALTH CARE SERVICES

LABOR ORGANIZATIONS

GOVERNMENT — FEDERAL, STATE AND LOCAL

Figure 5.1. Wellness Council.

Councils are adopting these guidelines. A graphic representation of the organization of the council is displayed in figure 5.2.

DELEGATE SELECTION: A KEY STEP

When a company joins the Wellness Council, the chief executive officer is asked to handpick a delegate and sometimes an associate delegate to represent the company at council functions. The delegates are generally from the company's personnel or human resources department, although other divisions of companies are well represented. For example, a company might send a health promotion professional, employee assistance coordinator, the fitness director, public relations director, or the sales manager. Sometimes the CEO represents the company.

The Wellness Council doesn't specify a term of office for delegates; that is left up to the company. Some delegates change because of transfers or promotions; others turn over the duties to the associate after a year or two of service. Some companies limit the term a delegate serves and periodically change delegates.

The important point is that delegates have a strong interest in health promotion and don't feel that this is just another committee assignment they have to fulfill as part of their job. Also important, the delegate should have direct access to the CEO and be visibly the CEO's

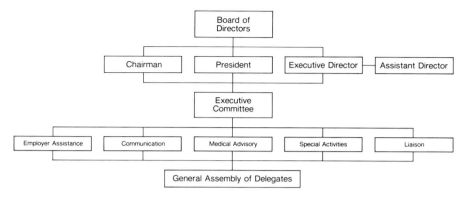

Figure 5.2. Organizational chart.

right-hand person on wellness matters. Mailings to CEOs with copies to delegates often encourage the CEO to contact the delegate for more information.

The CEO who elects to appoint someone to represent the company at delegate meetings of the Wellness Council will fall into one of three categories: The CEO who is a true believer in wellness will find another true believer within the company to become the delegate. This is the best working relationship of all. The delegate may be the systems programmer who likes to jog or the administrative assistant in personnel who plays tennis. Chances are, this delegate will be enthusiastic enough to bring back good ideas from Wellness Council meetings and, with the strong support of a CEO who is also committed to wellness, get exciting programs underway at the worksite.

Other CEOs may be interested in having the company join the council but may assign someone to be the delegate who is swamped with company work. Although this person has support from the top and goes to the meetings, he or she may follow through at a lower energy level. Yet another scenario could involve the skeptical CEO who thinks the company should join the Wellness Council, for the same reasons it joins the arts council, and assigns the duties of delegate to someone in, say, the personnel department. The assignee is not particularly enthusiastic about wellness activities either. He or she simply goes through the motions of attending meetings and looks at the whole idea as good public relations for the company.

Certainly the first scenario is ideal, but the point is that a company gets out of the Wellness Council what it puts into it. A half-hearted attempt to participate, because "it's the thing to do," is not going to benefit either the company's wellness programs or the Wellness Council.

There is yet a fourth consideration when appointing delegates. A

CEO, for whatever political reasons, may ask a high-level vice-president or senior manager to be the official delegate. But the CEO would be wise to appoint an associate delegate (the true believer) to do the actual work. In this case, the delegate is more a figurehead and a definer of wellness for the company, and the associate delegate attends the meetings and chairs the internal committees.

The original mission of the Wellness Council, which dates back to its formation, is to get the support of the CEOs. This continues to be a key strategy for any Wellness Council. Once a year the CEOs and their spouses are invited to a dinner and to hear an outstanding guest speaker on a health-related topic. Such a gathering serves to keep the good faith of the CEOs who have been supportive of the Wellness Council throughout the year and to remind them of the activities the council is involved in, especially when it comes time for the CEO to renew the company's membership and pay the yearly dues. Although the delegates from each member company do the work, the Wellness Council pays close attention to the CEOs, and they, in turn, are generally proud of their companies' involvement in the Wellness Council.

What Delegates Do

The delegate links the Wellness Council, the CEO, and the employees. Delegates show their active participation by attending the meetings, serving on a committee, reporting back to the CEO and co-workers, and initiating wellness programs at work.

Delegates and associate delegates should maintain an open line of communication with the CEO. If the delegate has direct access to the CEO on health promotion matters, such a link will make for a strong start within the company and for the council. Information flows from the top—as did the impetus to begin the company's wellness activities—and the support of the CEO should be carefully nurtured every step of the way.

Delegates are asked to attend delegate meetings which are held every two months—that's six lunches per year—and to hear a guest speaker. The Wellness Council office sends reminder notices (see figure 5.3) to delegates. The delegate assembly meets from 11:30 a.m. to 1:30 p.m., usually in a meeting facility provided by a member company.

One secret to WELCOM's success has been the decision not to meet too often. This is important in dealing with business groups. The meetings must, however, be quality meetings with good speakers, good materials to support the speakers' messages, and a nutritious lunch. As delegates begin to gather each time, they are often heard asking, "What's for lunch today?" Care is taken to be sure meetings end on time. On a few occasions, speakers have been politely interrupted. Business people appreciate such respect for their own schedules.

WELCOM

The Wellness Council of the Midlands

MEMO

DATE: September 16, 1985

TO: Delegates, Associate Delegates, Board of Directors

FROM: Harold Kahler, Executive Director

NOTICE OF MEETING
WELLNESS COUNCIL DELEGATES' MEETING
AT
NORTHWESTERN BELL
1314 DOUGLAS ON-THE-MALL
11:30 a.m. - 1:30 p.m.
FRIDAY, OCTOBER 4, 1985

TOPIC: Feelin' Good for Children

SPEAKER: Debbie Dodson, Coordinator, Feelin' Good for
Children, Fitness Finders, Inc.

Please return the registration form by **SEPTEMBER 27, 1985,** to:

WELCOM
1301 Harney Street
Omaha, Ne.   68102

WELCOM Delegates' Meeting
Friday, October 4, 1985

( )   I WILL ATTEND AND WISH TO BRING:

_____and_____

( ) MY CHECK FOR $7.00 PER PERSON IS ENCLOSED.

( )   I WILL NOT BE ABLE TO ATTEND.

NAME:_____COMPANY:_____

1301 Harney Street • Omaha, Nebraska 68102 • Telephone (402) 346-8962

The Wellness Council of the Midlands exists to promote wellness programs at the worksite. In pursuit of this mission the Council supplies the employer with a clearinghouse of information on worksite programs, offers the work force a source of support for achieving healthy lifestyles, provides the community with a positive environment for employer efforts to reduce health hazards and health care costs.

Figure 5.3. Reminder notices.

When the entire council meets, six times a year, it is easier to get active participation on the various committees and working groups to be described later in this chapter. Quality meetings are always more productive than a lot of boring programs or "busy work" committee meetings.

Member firms take turns hosting the assembly. Only the larger companies—with meeting facilities and dining arrangements—can accommodate such a large group because sometimes the number of delegates, alternates, and guests attending can be as high as 150 or more.

Figure 5.3. Continued

The Agenda

The delegates eat first. Special attention is given to menu planning by a registered dietitian to assure a well-balanced and healthy meal. The menu—printed on a tent card and pointing out items that are high in fiber and low in fat along with calorie counts—sits on each table. Typically, the meeting opens with announcements about national and local activities such as upcoming health fairs and corporate runs. A guest speaker then addresses the group.

Speakers are sometimes local experts, but as the council's budget increases along with its membership, the council is able to bring in guest speakers from around the country for the delegate meetings. One of the first speakers was Omaha's own nationally known stress expert, Dr. Robert Eliot, now at the Heart Lung Center in Phoenix. He is a cardiologist and is the author of *Is It Worth Dying For? A Self-Assessment Program to Make Stress Work for You, Not Against You.*

Here are some other examples of local and national speakers at delegate assemblies:

  Dr. Robert P. Heaney from Creighton University and Joan Werblow, executive director of the Dairy Council of Central States, on the topic of osteoporosis

  Richard Bellingham and Fred Roh from Possibilities, Inc., a consulting firm that helped design AT & T's health promotion program, on corporate cultures that support health promotion programs

Dr. Stan Haugland, director of a rehabilitation center at the Iowa Methodist Medical Center, on treatment for alcoholism and drug abuse

Debby Dodson Drake representing Feelin' Good, an aerobic program for children

Dr. Charles Berry, former medical director for the NASA space program and now a private consultant on health promotion, on the topic of smoking prevention and cessation at the worksite

Dr. Andrew J. J. Brennan, Metropolitan Life Insurance Company, company-sponsored health maintenance programs

Dr. Calvin F. Fuhrmann, chief of the respiratory division at South Baltimore General Hospital, on worksite smoking cessation programs

Bob King, a registered physical therapist and program director at Active Lifestyles in Omaha, on back care

Michael P. O'Donnell, director of health promotion at Beaumont Hospital in Michigan, on how to design worksite health promotion programs

Kris Berg and Dick Flynn, professors at the University of Nebraska at Omaha in health, physical education, and recreation, on the topic of fitness for living

The speaker is the final item on the agenda unless the chairmen of the standing committees have reports, which they make to the entire group only once a year.

THE ANNUAL MEETING

In November of each year, WELCOM holds its annual meeting for delegates, but the public is invited to attend this session. Such an open invitation has attracted health care practitioners, representatives from companies that are interested in joining the council and in health promotion, though they are not yet members; guests of delegates; employees from member companies; and the regular delegates.

This meeting is different from a regular delegate assembly because it is an all-day session that includes guest speakers, workshops, exhibits, round-table discussions, and informal meetings for networking. A prominent keynote speaker is a big drawing card for this meeting. For example, in 1985 the speaker was Dr. Robert Arnot, medical expert on the CBS Morning News. Dr. Arnot addressed the issues of the advantage of fit employees to corporations and the sudden death of fit people during exercise. Previous speakers have been Dr. Edward Diethrich, cardiologist and president of the Arizona Heart Institute, and Jess Bell, CEO of Bonne Bell Cosmetics and himself an enthusiastic runner who sponsors 10K runs for women throughout the United States.

Interestingly, Dr. Diethrich was two slides away from the end of his talk on coronary risk factors when a man in the audience collapsed. Dr. Diethrich dashed from the podium to the man's assistance—as did WELCOM's chairman, Dr. Zweiback, a surgeon. The victim, an employee of the local transit authority, suffered a heart attack-like reaction from the combination of sitting in a warm room and watching the slides of open-heart surgery but did not actually have a heart attack.

Attendees elect to participate in one of three tracks of workshops: One is for companies just beginning a worksite program. This is especially directed toward small businesses just getting started. Another is for companies with established worksite programs, and a third deals with individual programs such as stress management, back care, and nutrition. These workshops run through the rest of the day and are led by local experts, although sometimes an outside expert comes in.

The workshops for companies just beginning to get a worksite program going might be on how to set up a smoking policy or how to get a wellness program off the ground with little money. The more advanced workshops address the legal liabilities of fitness programs, how to motivate employees, and how to evaluate ongoing programs.

The informational brochure about the annual meeting serves as a registration form. Figure 5.4 is a sample.

The real business is conducted in the workshops, which give participants hands-on experience in health promotion at work. The idea is for them to come away excited but also with materials and ways to make programs work at the worksite.

In addition to council functions, delegates need time to devote to internal wellness activities. At a minimum, the actual time a delegate is involved in council activities averages out to only an hour or two per week. For example, reading communications, answering surveys, attending meetings, telephoning, and other routine activities do not demand a great deal of time. Of course, the more involved a delegate becomes in council activities, the more time will be required. Committee and task force assignments require more time. Even these activities, however, which all delegates should be encouraged to participate in, do not have overwhelming time demands.

Delegates may serve on one of several working committees and on an active board of directors.

Agenda

7:30- 8:30	**Registration**
	Early Bird Session — Body Composition Analysis will be conducted using the Bioelectrical Impedance Technique with interpretation at 12:30 p.m.
8:30- 8:45	**Welcome**
	Chairman, WELCOM — Eugene "Speedy" Zweiback, M.D. Surgical Associates
	President, WELCOM — Fred W. Schott Vice President, Training and Development Central States Health & Life Co of Omaha
8:45- 9:45	**Keynote: "Corporate Cost Savings With Fit Employees"** Robert Arnot, M.D. CBS Morning News
9:45-10:00	**Health Break** Mat Balcetis, M.S. Psychotherapist Therapist Supervisor Immanuel Medical Center
10:00-11:30	**Concurrent Sessions**
11:30-12:30	**Lunch**
12:30-12:45	**Body Composition Analysis Interpretation** Stephanie Anderson Alpha Fitness Center
12:45- 1:45	**"The Anomaly of Sudden Death in Fit People During Exercise"** Robert Arnot, M.D
1:45- 2:00	**Health Break**
2:00- 3:30	**Concurrent Sessions**
3:30- 4:30	**Wine and Cheese Reception**

CONCURRENT SESSIONS 10:00-11:30

A Legal Liability Issues
Michael A. Fortune, J.D., Erickson & Sederstrom Law Office, Omaha, NE.

This session will explore legal issues employers face when conducting wellness programs at the worksite.

B Double Participation, How We Did It!
Jackie Patterson, Director of Health Promotion and Wellness, Orlando Regional Medical Center, Orlando, Fl.

Find ways to double participation, spend less time to get more, use dollars better and share resources.

C Panel Discussion on Issues Facing the Health Care Industry
Representatives from local HMO, PPO, hospital administration, medical profession, Omaha Cost Containment Coalition and the insurance industry will discuss key issues in health care service

CONCURRENT SESSIONS 2:00-3:30

D "I'm in Charge"
Ruth Byers, Director, Health Systems Department, HNG-InterNorth, Omaha, NE.

This session will include ideas on how to help raise employee awareness, encourage worksite and personal wellness, and support wise health care consumer behavior

E Human Resource Management: Using Employee Involvement to Reduce Accidents and Absenteeism
Robert Jacobson, CEO Safeway Bakery Division, Clackamas, Oregon.

Sandy Humphreys, Corporate Health and Fitness Consultant, Clackamas, Oregon.

This session will explain the techniques used at Safeway to dramatically reduce accidents, absenteeism and their related costs.

F Carousel of Companies
The focus of this session will be an exchange of information concerning wellness programs and services provided by companies and agencies to their employees.

Registration

*WELCOM TO WELLNESS third annual seminar is held as an ongoing effort to provide the working community with positive educational information concerning company-sponsored wellness programs. In an effort to fulfill this goal, WELCOM TO WELLNESS will offer a variety of concurrent sessions with informative guest speakers to update seminar participants on current health promotion programs. We urge each company to register **two or more** wellness-oriented people to take an active part in the third annual WELCOM TO WELLNESS seminar.*

PLEASE TYPE OR PRINT

Name _____
       Last   First   Middle

Title _____

Business Phone: _____

Organization _____

Mailing Address: _____

PLEASE INDICATE YOUR PRE-CONFERENCE PREFERENCE(s):

Concurrent Sessions 10:00 - 11:30	Concurrent Sessions 2:00 - 3:30
A ☐	D ☐
B ☐	E ☐
C ☐	F ☐

FEES: *(Lunch included)*
☐ Members — $30.00 per person
☐ Non-Members — $40.00 per person

Please make your check payable to
The Wellness Council of the Midlands

Mail registration for each participant, along with a check to:
    WELCOM
    1301 Harney
    Omaha, NE 68102

REGISTRATION DEADLINE:
Friday, November 8, 1985. Facilities are limited. Early registration is recommended.

Figure 5.4. Registration brochure for annual meeting.

BOARD OF DIRECTORS

Each year, the delegates elect four members from within their ranks to serve on a 12-person board of directors. The primary responsibility of the board is to supervise the overall business of the Wellness Council.

As with all guidelines, some were made to be broken, and so it is with choosing board members. For example, the nominating committee (made up of delegates chosen by the chairman of the board of directors) might search the community to recruit a member to complement the make-up of the board. A suitable person from the community—say, a distinguished faculty member from a local university—might want to actively participate but be foreclosed because the university hasn't joined the Wellness Council yet. Bending bendable rules enables a council to include, rather than exclude, supporters it desperately needs.

Delegates with the most clout, visibility, prestige, and credibility should be handpicked to serve on the board of directors. A strong board sets the tone for the council and swings a lot of weight in the business community. After all, the boards of directors of the community's most powerful companies are made up of the most prestigious members of the business community.

Generally, the delegates unanimously accept the nominees suggested by an ad hoc committee. The board members are installed immediately.

The chairman of the board is elected by the board members. It is usually someone who has been serving on the board, not a new member. There is no term of office for the chairman, but the rule of thumb, for WELCOM, has been that the chairman serves two or three years or more. The bylaws are purposely not specific, thus giving the Wellness Council more options.

When the full 12-member board meets, members of the executive committee (which will be explained in the section on standing committees) are invited. Only board members have a vote, however; although the executive committee members can freely contribute to the discussions.

The board meetings are held a week or two prior to each delegate assembly (every two months) and at other times when necessary. The executive director, the chairman of the board, and the president of the Wellness Council put together the agenda—a combination of standard meeting fare, new issues, and solicitation of new members.

STANDING COMMITTEES

Five standing committees carry out the mandates of the board of directors. Delegates serve on committees of their choice based on their interests and special talents. The following is a discussion of each of those committees, detailing who is eligible to serve on them, how the members are selected, and what the committees do for the Wellness Council.

Medical Advisory Committee

Local physicians, dietitians, and professors from area universities and medical centers serve on the medical advisory committee. The committee reviews all proposed Wellness Council activities and prospective members to make sure that programs, activities, and materials are medically and scientifically sound. For example, to ensure that area health promotion service providers (such as the American Red Cross, the local hospitals, and consultants) have qualified programs, the committee screens each provider and publishes the results for council members.

Robert Murphy, a pediatrician and the first chairman of the medical advisory committee, sums up the reason why the activities of the Wellness Council must pass medical muster: "Nothing could undermine and shatter a wellness image faster than promotion of or endorsement of a medical wellness fraud to companies and employees" (personal letter to the author). The medical advisory committee is composed of professionals who can critically review information for scientific accuracy. Such a thorough review provides the member companies with a readily available source of accurate information.

Nutrition is one topic that is often portrayed inaccurately, and fad diets and diet plans are misleading and sometimes dangerous, says Ann Grandjean, a registered dietitian and member of the medical advisory committee. If a diet center wants a member of the Wellness Council to sponsor a workshop on weight loss for employees, the program the diet center is offering must be reviewed for medical soundness by the medical advisory committee before the committee will recommend the diet center as a provider.

Sometimes the medical advisers help delegates interpret technical subjects. During an annual meeting of WELCOM, for example, delegates were invited to take part in a demonstration of how to measure body fat. A local fitness center sent a registered nurse to hook people up to a bioelectrical impedance machine. By attaching electrodes to a person's foot and hand, a qualified technician can measure the percentage of body fat and interpret the results.

Although electrical impedance is one way to measure body fat, the medical advisory committee wanted to make sure that delegates knew the other more accurate methods of measuring body fat. So Kris Berg, professor of exercise physiology and committee member, told the group about the underwater weighing method and the use of calipers.

The medical advisers often give unselfishly of their time to put back into the community some of the support the community has given to them—beyond the care of their own patients. A host of consultants stands ready as a back-up resource for the medical advisory committee. These include professionals in sports medicine, dentistry, law, nursing, stress management, nutrition, and chemical abuse. The committee has the support of the Metropolitan Omaha Medical Society, and future councils would be wise to seek early support from the medical community.

Since it began in 1982, WELCOM has had as its major focus the establishment of corporate smoking policies and the general cessation of smoking at the workplace. To this end, the medical advisers continue to play a vital role in urging member companies to comply voluntarily with Nebraska's Clean Indoor Air Act. Basically, such compliance assures that members of the council have corporate smoking policies in which nonsmoking is the norm, not the exception. In fact, WELCOM initiated a major effort to help member companies address the smoking issue. Under WELCOM's name, counter plaques and window signs (figure 5.5) with the following message were printed and distributed: "An active member of the Wellness Council of the Midlands, this company respects the rights of the smoker and the non-smoker."

Communications Committee

Delegates who are professionals in public relations, publications, and media serve on this committee. Their expertise in publicity is important in making the general public and member firms and their employees aware of the council and its activities. They write press releases, screen publications, and advise on the development and placement of public service announcements.

The communications committee of WELCOM produced twelve 30-second public service announcements (PSAs). They were distributed by the council to area television stations. Although the Wellness Council provided the PSAs free, it asked cooperating stations to give an accounting of the number run and the amount of free air time. A grant from a local foundation covered the cost of production of the PSAs, which were given over $237,000 of free air time in Omaha.

The committee also publishes the council's *Executive Newsletter*, which is distributed four times a year to CEOs, delegates, and key decision makers. The newsletter is explained in more detail in the section on promotional materials.

WELCOM is discussing plans for publishing a newsletter for employees (over 70,000 are represented in WELCOM's total membership) to be sent to them at their homes or at work. Issues of cost and mailing list

An active member of the Wellness Council of the Midlands, this company respects the rights of the smoker and the non-smoker.

SMOKING IS PERMITTED IN DESIGNATED AREAS OF THE BUILDING.

The Wellness Council of the Midlands

Figure 5.5. WELCOM's smoking message.

maintenance are big obstacles to overcome, but a thriving Wellness Council can consider making the leap into direct contact with the thousands of employees and their families.

It's easy to take advantage of local avenues for free public relations. For example, local newspapers often print notices of meetings. In Omaha, a regional business journal routinely publishes a special edition devoted entirely to worksite wellness, and this has been a most beneficial public relations tool because the weekly journal is distributed to key business leaders.

Liaison Committee

Any delegate with expertise in any area may serve on this committee, which keeps up-to-date on the services and service providers in the community. Committee members then report this information to member companies. An example of a major undertaking by the medical advisory committee, the communications committee, and the liaison committee is WELCOM's Guide to Health Promotion Services.

The publication is a 48-page guide to service providers in the Omaha area. Member companies use the guide to find sources for information on the following areas:

Accident prevention and first aid	Hospitals
Alcohol and drugs	Mental health
Chronic diseases	Nutrition
Communicable diseases	Occupational wellness
Exercise and fitness	Poison control
Family help programs	Pregnancy and infant care
General health	Senior citizen health
The Handicapped	Smoking cessation
Heart disease and high blood pressure	Stress management
CPR	Weight management

A company, for example, that wanted to look into a stress management program for its employees could use the guide. Under stress management are listed 12 local service providers offering a wide range of programs. The Jewish Community Center, for instance, provides speakers, training, counseling, and health care in stress management. Other providers of stress programs are hospitals and nonprofit health organizations such as the American Red Cross. Figure 5.6 shows a sample page of this guide.

The guide is a first step in the decision-making process when a company is seeking an outside provider. The WELCOM stamp of approval on the programs is invaluable. Are all providers approved? No. A good example of those not approved are some of the faddish diet programs. But WELCOM doesn't publish a list of programs that aren't approved. Not being included is indication enough that the programs didn't pass muster.

The Wellness Council avoids endorsing a particular program and does no active programming of its own. By functioning solely as a clearinghouse, WELCOM is able to summon all the resources of the community and make them available to members. There is nothing competitive or controversial about a Wellness Council. It doesn't compete with any agency, public or private, whose activities relate to wellness. What the Wellness Council does do, however, is enhance the business climate for YMCAs or nutrition clinics or other agencies—profit and nonprofit. Member companies then make their own choices from a published inventory of approved programs.

Of course, publication of such a survey is time-consuming and tedious for all committee members, and the joint efforts of the three committees made the task a lot less overwhelming. By prescreening programs for medical soundness and suitability and range of offerings, the committees have simplified the search process for member companies.

Figure 5.6. WELCOM's guide to health promotion services.

Employer Assistance Committee

Delegates who serve on this committee assist member companies in selecting appropriate worksite wellness programs and help them keep the programs going. Delegates can form a consultation team that actually visits the worksites and makes suggestions to employers.

In addition, the committee is designing procedures for all member companies to use in evaluating their programs. Companies always want to know if the programs are actually saving health care dollars or boosting morale. But often, companies don't know how to measure the effectiveness of their programs. The employer assistance committee can give that expertise.

STRESS MANAGEMENT

"Never hurry, take plenty of exercise, always be cheerful and take all the sleep you need, and you may expect to be well."
J. F. Clarke

American Red Cross
3838 Dewey Avenue
Omaha, NE 68105
341-2723
SP   TR   TH

Bergan Mercy Hospital
7500 Mercy Road
Omaha, NE 68124
398-6862
BK   SP   TR

Dept. of Preventive and Stress Medicine
(Stress Health Physical Evaluation Program - SHAPE)
42nd & Dewey
Omaha, NE 68105
559-4112
TH   HC

Immanuel Medical Center
6901 N. 72nd St.
Omaha, NE 68122
572-2121
BK   SP   TR   TC

Jewish Community Center
333 South 132nd St.
Omaha, NE 68154
334-8200
SP   TR   TH   HC

Lutheran Family and Social Service of Nebraska, Inc.
120 South 24th St.
Omaha, NE 68102
342-7007
BK   TR

Lutheran Health Services
P.O. Box 3434
515 S. 26th Street
Omaha, NE 68103
536-6790
TR   TH   HC

Mercy Hospital
800 Mercy Drive
Council Bluffs, IA 51501
328-5000
BK   FL   SP   TR   TH   HC

Methodist Hospital
8303 Dodge
Omaha, NE 68123
390-4526
SP   TR

University of Nebraska Medical Center
42nd & Dewey
Omaha, NE 68105
559-4152
TR   SP

Upjohn Health Care Services
8031 West Center Road, Suite 226
Omaha, NE 68124
392-0600
BK   TR

Visiting Nurse Association
10840 Harney Circle
Omaha, NE 68154
334-1820
BK   TR

BK = Booklets and Pamphlets
FL = Films
SP = Speakers
TR = Training/Classes

TH = Therapy/Counseling
FN = Financial Aid
HC = Health Care
SR = Screening

Figure 5.6. Continued. A sample page.

The committee's other activities include building a library of resource materials at the Wellness Council office or through the reserve desk at the local public library or library of a medical center.

Special Activities Committee

Members of this committee plan and coordinate unique community-wide activities that will motivate people to adopt more healthy lifestyles. This is the committee for delegates who have working experience in conference planning and organizing events. Each year the committee plans and carries out the annual November meeting. It is also involved in special projects such as the Corporate Wellness Series, which are activities with special emphasis on family participation and educational clinics. Fitness events—a run, a swim, and biking and golf events—are part of the wellness series. It's a participation event, not a health fair.

For the annual meeting, this committee picks a guest speaker, decides the topics of the workshops and round-table discussions, contacts vendors of health services, and invites them to set up displays in the lobby of the hotel or conference center where the meeting is held. Members handle registration, put together information packets, design the registration form, and evaluate the success of the meeting. Results of the evaluation are incorporated into the program for the following year.

A first-time event—HealthTask '86, a national conference on worksite wellness—was co-sponsored by the Wellness Council of the Midlands and the Health Insurance Association of America in June 1986. The special activities committee coordinated the national effort which attracted over 500 participants to the conference in Omaha.

Such national gatherings are a natural extension of the efforts of Wellness Councils as the movement spreads to cities everywhere. Planners envision statewide and regional conferences as well.

Executive Committee

The chairmen of each of the standing committees serve on the executive committee along with the chairman of the board of directors and the president of the Wellness Council. The presidential position is customarily filled by a delegate who is handpicked by the chairman of the board to be the active doer—the emcee of the delegate meetings, for example. As discussed earlier, the executive committee is invited to meet with the board.

Each year the president of the Wellness Council appoints a nominating committee. The committee, along with nominating board members and officers for election, selects delegates to chair the committees. The chairmen are then formally appointed by the council president. The input of the executive director is vital to this process. As in all organizations, regardless of formal rules, strong direction by the professionals insures good choices for important committee positions.

PROFESSIONAL STAFF

Executive Director

The executive director is responsible for the administrative functions of the Wellness Council. Any council would be wise to bring in an executive director even in the formation stage of the council. Full job descriptions for this position and for the assistant director appear in appendix B. The board of directors screens applicants and hires the executive director who in turn hires the assistant director.

The best combination of skills an executive director could have would include strong interpersonal communication skills because the job basically involves a lot of public relations. Good organizational skills are also a plus because the executive director keeps in contact with hundreds of delegates, community health care providers, and CEOs. Administrative duties involve keeping a close eye on the budget, scheduling meetings, and creating publications. The third asset is good public-speaking skills. The executive director is called on to make presentations to large groups such as the Rotary Club, the State Department of Public Health, professional meetings, and to small groups such as company wellness committees and even one-on-one meetings with CEOs and other decision makers. Enthusiasm from the executive director sets the tone; it is essential.

Because the field is so new and so untested, finding the perfect person to fill the executive director's position is difficult. Some colleges and universities are adding programs in fitness management and health promotion, but graduates of these are often young and inexperienced, although they may have the technical knowledge. Prospective candidates may come from the fields of public health administration, nonprofit agencies, foundation management, or health promotion.

The salary range to draw such a qualified candidate certainly depends on what is available in the budget, the size of the community, the cost of living, and other unknowns. WELCOM suggests a salary range comparable at least to that of middle managers.

The hiring of an executive director can be difficult in the early years because of budgetary restrictions. However, there are a few options available even to those councils with limited initial budgets. For example, the executive director can be an employee of one of the major sponsors of the council, and the salary be paid by that sponsor. Or the major sponsors may want to each commit a certain amount of money over an agreed upon period of time to fill a paid position. Another option would be to have the executive director be an employee of one of the major sponsors, but have that person devote a portion of his or her time, say 30 to 40 percent, to the duties of the Wellness Council.

A third option is to have one of the local businesses loan an executive director to the council. The person may be part-time or full-time, but in either case the salary is paid by the local business. This is a common practice among businesses especially among those making executive loans to nonprofit organizations such as the United Way. A fourth option is to hire a retired business executive whose salary can be waived or negotiated.

Fast-growing Wellness Councils might hire an assistant director who serves as a part-time secretary and who stands in for the executive director at committee meetings. Sometimes arrangements can be made with local universities to provide interns from the departments of exercise science, communications, and public relations. These college students can organize reference materials, answer phones, run errands, and organize events.

Office Space

Not long after WELCOM's successful beginning, it became obvious to both the Wellness Council and the Omaha Chamber of Commerce that their purposes were each served by the other. Representatives of the Omaha chamber said that wellness fits nicely into their 10-year plan and noted that worksite wellness was, in fact, predicted to be a strong national trend. Omaha is fortunate to have a strong and active chamber, and WELCOM was eager to lease office space in the chamber's new building in revitalized downtown Omaha. It has proven to be a beneficial relationship for both WELCOM and the chamber. Such natural coalitions should be sought with local chambers of commerce and with other groups with compatible goals such as cost-containment coalitions, health information networks, health and wellness consortiums, and governors' committees on health promotion.

The Wellness Council contributes a regular column to the Chamber of Commerce newsletter. This is a specific way to reach small businesses and to help them become familiar with health promotion at the small business worksite. The column is written by WELCOM members in small businesses that belong to the Wellness Council.

PROMOTIONAL MATERIALS

WELCOM publishes a number of printed materials as part of its mandate to provide information and be a clearinghouse for information. Following is a discussion of some of these publications.

Payroll Stuffers

WELCOM makes available to member companies a series of 24 inserts that companies put inside payroll-check envelopes. Some companies distribute them directly to employees or republish the text in company newsletters. Others stack them on reception tables which are open to employees and clients who happen to pass by and are attracted to the colorful little flyers. Current topics are

Back care	First aid and choking
Colorectal cancer	Food selection
CPR	High blood pressure
Diabetes	Immunization
Dietary guidelines	Risk factors
Driving and your health	Skin cancer
Exercise	Smoking
Eye safety	Stress management
First aid	Water

The payroll stuffers are developed in the following way. The executive director may see a need for an informational handout on risk factors. The director then seeks a local resource, which, in WELCOM's case, was the American Heart Association—Nebraska Affiliate. The as-

Figure 5.7. Payroll stuffers.

sociation wrote the text, and WELCOM's medical advisory committee approved it. The association's endorsement of the text is important; in fact, its logo and name appear on the cover of each next to the WELCOM logo. Each payroll stuffer has a different sponsor (for example, the American Red Cross was involved in producing the stuffer on first aid, and the National Society to Prevent Blindness, Nebraska Affiliate, supplied information for the flyer on eye safety).

Print runs of 30,000 to 50,000 assured that enough stuffers would be available for member companies. Member companies then pay one cent per stuffer so that WELCOM can recover some of the outside costs.

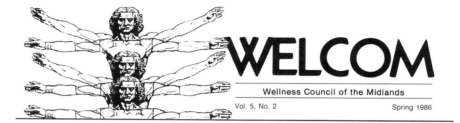

WELCOM

Wellness Council of the Midlands

Vol. 5, No. 2   Spring 1986

College President Stresses Developing The Whole Person

Dr. Kenneth R. Nielsen

College of Saint Mary in Omaha is one of the newest members to join the Wellness Council of the Midlands.

"I believe that being able to work with different lifestyles is a key to developing a quality wellness program," said Dr. Kenneth R. Nielsen, president of the college. "The lifestyle of the American population is always changing, and keeping abreast of their needs is important in a wellness program, as with the college, especially in our curriculum offerings. Understanding a person's need has enabled the college to offer programs that are flexible, marketable and designed for consumer convenience.

Nielsen assumed the presidency of the college in June 1984. He came to Omaha from Seattle University where he was a vice-president.

Nielsen's support of the wellness concept can be seen in his personal interest in physical fitness. He has been an avid runner for 27 years. Nielsen stated that he has run around the world once if you calculated all the miles he has run over the years. A major setback to his running happened last year when he had to undergo surgery on his back. Having made fitness part of his lifestyle, he continues with daily exercise limiting himself to certain programs.

The physical as well as the mental aspects are important to Nielsen. At the college he stresses developing the whole person.

The college presently has a variety of activities for employees and students. Health awareness programs are offered in the dorms. The college cafeteria makes an effort to provide information on each entree item, while providing healthier eating tips. Students, faculty and administration are encouraged to participate in the campus health awareness week and in other organized community events such as the Corporate Cup.

The college is a developing, private, independent Catholic institution of higher education, offering an integrated liberal arts component with professional and career preparation in both the associate and bachelor degree programs. Enrollment is about 1,100.

"Through joining the Wellness Council of the Midlands I am aiming at providing the college community an opportunity for more than just designated health weeks," said Nielsen. "When looking at our different activities over the year I see WELCOM as a great source of new information for our institution. We desire to overcome our negative habitual patterns with accurate, usable knowledge about the way in which we live our lives. We are looking forward to joining the corporate community in working toward these goals."

WELCOM welcomes the College of Saint Mary. □

GM Chairman to Keynote the National Conference in June

Roger Smith

Roger Smith, chairman of General Motors, known for his own pioneering efforts to bring worksite wellness to the thousands of members of the United Auto Workers union, will speak at HealthTask '86.

The conference, sponsored jointly by WELCOM and the Health Insurance Association of America, will be held from June 16 to 18 at the Red Lion Inn in Omaha.

A wide-ranging and comprehensive program of speakers and dialogue sessions is taking shape for the conference under the direction of a program committee headed by Fred Schott, WELCOM's past president.

"HealthTask '86—Working for Wellness" is the theme for the three-day meeting whose purpose will be to sharpen management skills in planning, conducting, and evaluating health promotion programs for employees of small and large businesses.

Several hundred business executives and health promotion specialists are expected to attend the conference which is also drawing top authorities such as Dr. Charles Berry, former medical director for NASA, and Dr. Jonathan Fielding from UCLA to lead the working sessions.

Contact WELCOM's office for registration information. □

Figure 5.8. Executive newsletter.

Members usually buy one for each employee. The entire series is always available, so new members can begin distributing any of the flyers. Established members continue to distribute new flyers as they are developed. A company that pays people twice a month, for example, could put a different flyer in each paycheck for one year.

Although this may seem like a lot of paper and a lot of personnel time spent in stuffing and distributing, this method of educating employees is relatively inexpensive, is a good way to start wellness awareness within a company, and says to employees, "the company cares enough about you to give you health information." Large companies sometimes have a policy not to include anything other than a paycheck or direct-deposit receipt in the pay envelope. This potential barrier to distributing educational information on health can be overcome if the delegate to the Wellness Council has the full support of the CEO.

Examples of payroll stuffers are reproduced in figure 5.7.

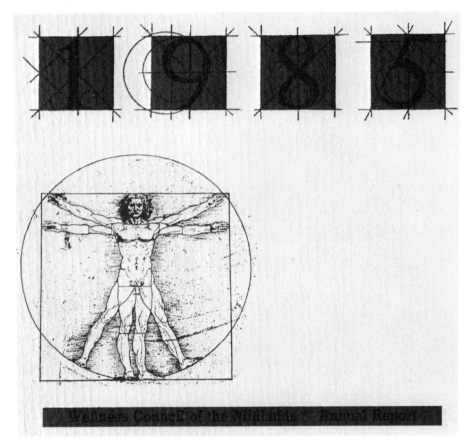

Figure 5.9. Annual report.

Newsletter

WELCOM's *Executive Newsletter* keeps CEOs, delegates, and other interested people informed about council activities, what member companies are doing, new members, committee reports, and national health promotion trends. The *Executive Newsletter* is published quarterly, and distribution has grown to nearly 4,000.

Courtesy copies are distributed to government agencies, trade associations, CEOs of some Fortune 500 companies, and health promotion organizations.

The assistant director of the Wellness Council generally takes responsibility for soliciting articles from member companies. This is often easy because public relations departments are eager to provide material on their companies. Committee chairmen submit brief reports

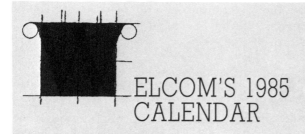

ELCOM'S 1985 CALENDAR

Roger Koehler, Vice President Education Services
Methodist Hospital
A health risk appraisal that was handed out and completed at the previous delegate meeting was reviewed and interpreted for the delegates.
Host: Immanuel Holling Center.

Fred Roh, M.A., Vice President, Possibilities, Inc.
Basking Ridge, N.J.
AT&T's Total Life Concept program was explained from the inception of the idea through the first evaluation process. Host: Methodist Hospital Continuing Education Center.

Debbie Drake, Program Coordinator
Fitness Finders, Inc., Spring Arbor, MI
Debbie Drake visited a local YMCA, spoke to teachers of the Omaha Public School System, and addressed the delegate meeting during her brief stay in Omaha. Host: Northwestern Bell.

Robert Arnot, M.D., CBS Morning News
Dr. Arnot addressed the participants on the anomaly of sudden death of fit people during exercise and the importance of healthy employees to the bottom line of corporations. Host: Peter Kiewit Conference Center.

■ JUNE 7, 1985
WELCOM
DELEGATE MEETING:
HEALTH RISK PROFILES

■ AUGUST 2, 1985
WELCOM
DELEGATE MEETING:
AT&T TOTAL LIFE CONCEPT

■ OCTOBER 4, 1985
WELCOM
DELEGATE MEETING:
FEELING GOOD
FOR CHILDREN PROGRAM

■ NOVEMBER 15, 1985
ANNUAL SEMINAR:
WELCOM TO WELLNESS III

5

Figure 5.9. Continued

also. Health promotion vendors often supply informative material which can be developed with a local angle. Short items and facts are reprinted from other health promotion publications. Figure 5.8 is an example of the *Executive Newsletter.*

At this time, the *Executive Newsletter,* distributed to senior-level management and delegates at each company, is printed on quality paper using black ink. Because it is in this camera-ready form, individual companies may clip articles to insert in their own company newsletters for distribution to all employees.

Annual Report

Technically, member companies are the stockholders, so the Wellness Council publishes an annual report to keep them informed on the year's activities. The 1985 report, for example, gave brief descriptions of programs in selected member companies and printed photos of CEOs and committee chairmen. The usual budget information and projections also appear (see figure 5.9).

STAGES: FOUNDATION, FORMATION, AND GROWTH

As Wellness Councils get off the ground, they will probably follow the same stages of development that WELCOM has. The ideas came in the foundation stage, which involved a lot of talking. At this point the concept was envisioned and the benefits were projected.

The talk turned into action, and the concept was designed and put into place. WELCOM was incorporated, the mission statement written, companies joined, committees and the board of directors were formed, and the delegates became active.

Once the council is past the foundation stage and into formation activities, it might be useful to poll members about what programs are in place and what their goals are for wellness. WELCOM and the Health Insurance Association of America have developed a survey form for councils to use. Delegates could be asked to fill out the form, and the council's executive committee can report the aggregated results at one of the first delegate meetings. The form is free of copyright restrictions and reprinted here as figure 5.10.

WELCOM has entered the third stage—growth. The fears and doubts about whether this organization would flourish are over. WELCOM is both financially and organizationally established. The emphasis now turns from "what shall we do?" to "how can we do it better?" Committee goals are reviewed and revised if necessary. Alternative methods of fund raising are being considered, and methods to enhance communications and fulfill the Wellness Council's mission are underway.

Membership Survey

Company Name _____

Address _____

       Zip_____ Telephone ( )_____

Name of CEO _____

Type of Business _____

Number of Employees_____ Salaried_____ Hourly_____

Age Distribution: 18-29_____ 30-45_____ 46-60_____ over 60_____

1. What is the present involvement of your company in providing opportunities for improved health and well-being for your employees and their families?
   _____
   _____

2. Please check the services/information you provide:

   Participatory Activities:
   ___ Aerobics
   ___ Blood Pressure Reading
   ___ CPR
   ___ Employee Assistance Programs
   ___ Exercise Opportunities
        (Club membership, walking
        groups, etc.)
   ___ Individual Health Appraisals
   ___ No-Smoking Classes
   ___ Sports Teams
   ___ Other: _____

   Information/Education Activities:
   ___ Alcohol/Drug Abuse
   ___ Cancer Awareness
   ___ Company Smoking Policy
   ___ Health Promotion Material
   ___ Incentives for Good Health/
        Absenteeism
   ___ Nutrition Awareness (Cafeteria,
        vending machines, etc.)
   ___ Stress Reduction
   ___ Weight Control

3. Are you currently investigating and/or do you wish to start any of the types of wellness programs listed in Question 2? Which of the above would be a benefit to your company?
   Please list: _____

4. What is your personal involvement in fitness activities?
   ___Walker   ___Runner   ___Health Club   ___Sports   ___None

5. Does your health insurance program provide incentives for self-health?
   _____
   Name of insurance company:_____

6. What is your goal in providing/encouraging wellness opportunities for your employees in the next year? _____
   _____

7. What services or assistance would you need or find helpful?
   ___Speakers
   ___Public media campaign on health
   ___Training for in-house wellness coordinator
   ___Program suggestions
   ___Consultant assistance for initiating program
   ___Information (payroll stuffers, printed materials, local program resources)
   ___Cost/benefit information
   ___Health promotion resources in community

8. What materials, facilities or other resources do you have available for a worksite wellness program? _____
   _____

9. Specialty staff in health/fitness:___Nurse___Physician___Wellness/fitness___Other

10. My name (if other than CEO) _____
    Position _____
    Address _____
           Zip_____ Telephone ( )_____
    ☐ Check here if you would like a report of the survey results.

Other comments: _____
_____
_____

Return to: _____

Figure 5.10. Membership survey.

More sophisticated events like the Corporate Wellness Series (swimming, biking, running, and golf) involve the community and bring family members into full participation. These events come with maturity as the Wellness Council, with cooperation from local businesses, shapes the health of the community.

KEY STRATEGIES

WELCOM's founders, whether by design or default, used a number of key strategies to make the model work in Omaha. Here they are:

- Gain the support of the CEOs by finding a respected business leader already committed to wellness to lead the effort. If that leader can then get several other key leaders to co-sponsor or endorse the effort, other business leaders will at least consider the concept of worksite wellness.
- For the kickoff luncheon, seek a good speaker with impeccable credentials.
- Prepare a well-designed packet of information, including endorsement letters, supporting the soundness of worksite wellness. Ask not only for financial support but for help and input in formalizing the council. This will enhance the general sense of corporate ownership.
- Allow delegates at the first delegate meeting to have meaningful input and involvement in setting up the council structure. Develop first year goals to create a real sense of action and commitment to the cause.
- Keep the cost of dues reasonable. Compare a modest $2 per employee per year against as much as $200 per employee per month for health care insurance. Seek foundation funding. Cooperate with government agencies but do not seek government funding.
- Cultivate natural allies such as the insurance industry. Their bottom line is directly affected by the high cost of health care too. Also form a bond with the local Chamber of Commerce, whose interests are indeed compatible with the Wellness Council's.
- Seek the support of local media.
- Keep the founding CEOs informed and as involved as they want to be.

PART THREE

6

THE INSURANCE INDUSTRY TAKES THE LEAD

Every viable business must continually ask itself, "What business are we really in?" Success—even survival—depends on the answer to that question. Examples of this abound, as John Naisbitt so aptly pointed out in *Megatrends*. AT & T asked years ago, "What business are we in?" and concluded that they were no longer in the telephone business but in the communications business. Many of our great railroads fell on hard times because they could not understand that they were fundamentally in the transportation business, of which railroads would become an increasingly smaller component. IBM and Xerox—two of our greatest corporations—by continually refining their answer to the question, "What business are we really in?" understood that they were not just in the typewriter and copy machine business but are, in fact, in the business of office automation.

Today, not just a few progressive insurance companies but the entire insurance industry is asking, "What business are we really in?" For years insurance companies were—and still are—in the business of protecting people from the financial tragedy that can accompany sickness, accidents, and death. However, health and life insurers are redefining their mission to include helping people improve their health and prolong their lives.

It's no secret that insurance companies are in business to make money. Historically, the role of insurance companies has been to spread the risk and insure it. They're also in the investment business: They sell policies, collect the premiums, invest the money, and strive for a good return on those investments.

But Dr. Arthur Ulene, medical consultant to network television, summed it up best in his remarks to the insurance industry's conference "Health Education and Promotion: Agenda for the 1980s":

> (Insurance companies) are also in the health business, because what you do influences our health policies and the health of our people. Your industry plays a great role in the incentives that motivate people to be healthy or to be sick. The crazy thing is that the insurance business pays people to be sick. You must have a disease in order to be reimbursed. So we make people sick so that you will pay for a visit. In fact, the only way for people to get a day off is to be sick. If you take a day off, you lose the money, but if you get sick, you get paid for it. Is it any wonder that people tend to be ill approximately the same number of days as they have sick time?

For far too long the insurance industry played the role of neutral bystander. But as actuaries continue to develop more precise ways of measuring how lifestyle and habits make a difference in how long people live and how often they get sick, the insurance industry will continue to step into the forefront by taking an active role in health promotion. The advisory council on health of the Health Insurance Association of America identified the worksite as the place where the industry should make its best effort.

Such revelations for health insurers opened up new options and led the way for them to develop new products that rewarded healthy lifestyles. Back in 1957, when Surgeon General L. E. Burney declared war on smoking and, later, when that familiar warning began to appear on cigarette packages, the insurance companies followed suit by lowering rates for nonsmokers.

These innovative insurance plans, in combination with active worksite health promotion, pay off. Dr. Ulene says it's time to stop letting the medicine tail wag the healthy dog. He foresees a system in which health promoters, not doctors, become the primary health professionals.

BE WELL, PAY LESS

Insurers are often taken to task for not doing something for the people who are healthy. Companies and consumers are looking for price breaks similar to those offered to people who have safe driving records. Why not give price breaks for people who live healthful lives? The industry is responding to this demand in two ways: (1) with group policies for companies that have health promotion programs; and (2) with individual health insurance for buyers who prove they are healthy.

Nearly 200 million Americans have private health coverage through insurance companies, Blue Cross/Blue Shield, HMOs, and other plans. More than 80 percent of these people are under group insurance plans through employers.

The insurance industry, through the example of some forward-looking companies, is beginning to reward client companies for their health promotion activities. Rewarding, in this sense, means that companies can pay lower rates for their health insurance in some cases.

One particularly innovative plan is offered through Northwestern National Life Insurance Company in Minneapolis. Because companies are moving in the direction of revamping current health care plans (cost shifting), and because employees are becoming even more responsible for being wise health care consumers and taking charge of their own lifestyles, this important new direction for group health insurance deserves a look.

Under Northwestern National's Alliance program, the employer benefits right away because the insurance company has set a high deductible (possibly $500 to $2,000 per family). The employer may then pay even less (perhaps 3 percent less) in following years if employees are healthier—in other words, if they use less expensive medical care, have fewer days off for health reasons, and show greater productivity.

To encourage employees to adopt healthier lifestyles, the insurance carrier and the company promote wellness in a number of ways. The overall health risk of the employee group is measured using a health risk assessment. Once a year, employees undergo a confidential health screening and complete a questionnaire covering personal and family health history and personal lifestyle. This health risk assessment serves two functions: (1) Each employee receives a confidential report outlining his or her personal potential health risks; and (2) the employer receives a report about the group's overall health. That report identifies what action the employer can take to prevent or modify major risk areas that are uncovered in the health risk assessment.

The insurance carrier then can help the employer set up worksite health promotion programs targeted to the company's particular health risks. For example, a company may have an unusually high number of smokers. The insurance company can introduce the company to a vendor or community resource that specializes in smoking cessation programs. By working together, the insurance company and the employer can minimize the risk for the insurance company, which then passes along those savings to the company by lowering group health insurance rates.

But what about the employee who is faced with paying such a high deductible? The Alliance program allows for a Flexible Spending Account, which is part of a cafeteria plan for fringe benefits. Flexible Spending Accounts are set up by the employer for each employee. Pretax dollars in the accounts are used to pay for eligible health-related expenses, including the deductible itself. At the end of the year, any money remaining in an employee's account can be paid to the employee as taxable wages. How many pre-tax dollars the employee contributes to the Flexible Spending Account, and how many pre-tax dollars the employer puts in, is determined by each company. Perhaps the company will split the cost with the employee.

There is no doubt that the company will save money up front with this revolutionary plan. The dollars so saved could easily be put into Flexible Spending Accounts. But the real saving is in employees' health. By educating employees with the health risk assessment, and making them more accountable for spending "their own money" on health care costs, employers can encourage employees to become smart consumers and become attuned to the idea that lifestyle change does indeed make a difference on the bottom line—both for the company and the employee.

What about the other 20 percent of the population—the millions of people who cannot qualify for group insurance because they do not belong to a group? They are the part-time workers or the self-employed—accountants, lawyers, ranchers, farmers, pharmacists, doctors, small business owners, and so on. Because they must buy individual major medical health insurance, they pay more.

For families and individuals, these health insurance premiums are generally based on age. If you're 35 years old, you pay what 35-year-old people across the country pay. It's like not being able to negotiate on the sticker price of a new car. And even more frustrating for these people, individual premium rates are higher for them because their individual risks are so unpredictable.

Central States Health & Life in Omaha now offers a Healthy American policy for which premiums are based not on chronological age, but on biomedical age (whether your body is as old or as young as the number of candles on your birthday cake). People leading healthy lifestyles would certainly pay less. And such a rating system could eventually affect all those millions of Americans who do not work within the corporate structure and who are not able to participate in health promotion programs at work.

SMOKING INITIATIVES

In addition to the efforts of individual insurance companies to improve health status, the Center for Corporate Public Involvement (jointly sponsored by the health and life insurance industries) initiated a major effort to get people to stop smoking.

Here are five reasons why the insurance industry chooses to play a leadership role in encouraging Americans to stop—or not to start—smoking:

1. The evidence of a direct link between smoking and the leading causes of death and disability is overwhelming.
2. The economic toll of smoking on our society is enormous.
3. Smoking is the single most important preventable cause of death.
4. Every time efforts are intensified to spotlight the risks of smoking, more smokers give up the habit. Surveys show that 90 percent of smokers have either tried to quit or would quit if they could find an effective way to do so. They need help.
5. Dollar for dollar, the insurance industry would achieve more by sponsoring smoking cessation programs than by supporting any other initiative to improve health and cut health care costs.

We all know the familiar statistics and even know someone whose life was cut short because of smoking. Since the first Surgeon General's report in 1957, some 30 million Americans have stopped smoking. Still, 55 million people continue to smoke. The time is overdue for the tobacco industry to exercise public accountability for the hazard of their product and to join society in its effort to protect, preserve, and promote the health of the people.

There has been disappointingly little progress, as L. E. Burney, Surgeon General (Ret.) pointed out:

In 1957, as Surgeon General of the Public Health Service, I called a televised press conference in Washington, D.C., to state and provide supporting scientific data that cigarette smoking was a major causative factor of lung cancer. This was a significant milestone in the condemnation of cigarettes as a health hazard. My statement was cleared by the White House and by the Secretary of Health, Education and Welfare giving it greater validity as to content and as the first public announcement by the primary health official of the National government.

A second statement in 1959 documented further evidence of the direct cause and effect between cigarette smoking and lung cancer and also presented scientific evidence implicating cigarettes as a major causative factor in heart disease, emphysema and other health conditions.

Now, 30 years later, what has happened? In 1956, the deaths in white males from lung cancer numbered 29,000. In 1985, some 350,000 Americans died prematurely from cigarette smoking.

Still, the tobacco industry and its Tobacco Institute reject the overwhelming scientific evidence today as they did almost thirty years ago, convicting cigarette smoking. They continue to spend millions of dollars to promote and encourage the consumption of cigarettes. Freedom and license are not the same thing (personal letter to the author).

Employers can't close their eyes to the devastating impact of smoking on the well-being of employees and to its effect on the bottom line. Smoking burns a $65 billion hole in America's economy every year.

Smokers are also policyowners. Health insurers have to ask themselves: How many claim payments must nonsmoking policyowners pay for the treatment of coronary disease, cancer, and respiratory diseases that might have been reduced or even avoided if the patient hadn't absorbed nicotine and carbon monoxide every time he or she lit up?

Surgeon General Koop calls smoking "slow-motion suicide" (*New York State Journal of Medicine*, 1983). And Dr. Ulene told the insurance industry, "Cigarette companies don't advertise for their health or ours—they advertise because those ads sell cigarettes. The same is true for the people who make Twinkies. The real question is what can we do to swing the balance in favor of health promotion?"

Still the battle continues to be played out in the media and between doctors and cigarette companies. Cigarette ads are not allowed on television. Should that ban include print media as well? The first amendment freedoms may determine that issue, but newspapers and magazines are free to reject cigarette advertising on their own. Some, like *Reader's Digest*, *The New Yorker*, and *Good Housekeeping*, already do.

Call it enlightened self-interest, if you will, but the insurance companies must first adopt effective programs to reduce the toll of smoking on their own employees. Then, after they have attained significant results, let them offer their expertise to those they serve: the insurance clients.

Already several prominent insurance companies have acted boldly and announced total bans on smoking in their buildings. These include Aid Association for Lutherans, Connecticut Mutual Life, Lincoln National Life, Mutual Service Life, Northwestern National Life, Provident Indemnity Life, and United Services Life. Others will surely follow.

KEEPING THE MOMENTUM GOING WITH WELLNESS COUNCILS EVERYWHERE

In 1984 the Health Insurance Association of America (HIAA) started the bandwagon rolling to develop Wellness Councils around the country. They did this by sponsoring a teleconference linking business and community leaders in 25 cities. Over 2,000 decision makers viewed the video conference that was transmitted around the country by satellite.

President Reagan and Margaret Heckler, former secretary of the Department of Health and Human Services, ignited the spark, and HIAA has capitalized on the momentum and visibility of bringing together so many interested business leaders.

Out of the teleconference grew a nationwide effort by the Health Insurance Association of America to create a network of Wellness Councils in cities across the United States. Why? Because Wellness Councils work. The Wellness Council of the Midlands serves as the model for these new initiatives.

Even more astounding is the acceptance of the idea of forming a national network, and even though the HIAA targeted only five cities to start councils in 1985, over 11 formed, and other communities continue to jump on the bandwagon.

The HIAA recognized the importance of backing Wellness Councils and provided $150,000 for three years to fund the Wellness Council program. HIAA envisioned the Wellness Councils of America (WELCOA) as the HIAA-sponsored umbrella over all local councils. A full-time staff person within HIAA coordinates the effort (to proselytize, cajole, convince, and show real dollar savings to CEOs in large insurance companies everywhere). The idea is to find a CEO in a major insurance company who will take the lead in each city.

Outside funding to support WELCOA came from the Office of Disease Prevention and Health Promotion under a cooperative agreement to support national health promotion programs. A grant of $10,000 was made to HIAA for the development of public service announcements on worksite wellness. The PSAs will encourage employers to sponsor wellness programs in their companies and will be distributed to national network television, local affiliates, and public broadcasting stations.

The chairman of the health education committee of HIAA, James R. Brennan, who is also vice-president of Northwestern National Life Insurance, addressed the Senate Subcommittee on Health in June 1985. Brennan summed up the trade association's and insurance industry's commitment to forming councils by stating, "It is an exciting prospect to envision the Omaha council replicated in other cities where private initiative can be put to work to improve health practices and reduce the social and economic toll of illness and disability in the community" (congressional testimony, 1985).

In October 1985, eleven Wellness Councils were awarded charters from the Department of Health and Human Services and HIAA. Among them was WELCOM, recognized as the first of many. Councils operating in Tucson, Ariz., and Norristown, Pa., were also recognized. And councils forming in Atlanta, Ga.; Chattanooga, Tenn.; Columbus, Oh.; Jacksonville, Fla.; Milwaukee, Wisc.; Minneapolis, Minn.; Greensboro, N.C., and Baton Rouge, La., were also given congressional recognition.

An insurance executive, Lynn Johnston, CEO of Life of Georgia, in his excitement about the corporate concern for wellness, told an industry group, "The American audience will accept this kind of message from insurers, thanks to our history of long association with concepts like this. It is far less likely to accept the wellness message from retailers, stockbrokers, and banks" (International Claim Association meeting, 1985).

7

WELLNESS IN THE YEAR 2000

Asbestos, noise pollution, toxic wastes—headlines warn about the dangers of disease and death in the workplace. Workers who were unaware of the dangers have been exposed to hazards in the environments where they have often worked for years. Millions of dollars are being spent in litigation of such cases. Precious resources, at a time when business can least afford it, are being spent in trying to correct the effects of the unhealthy workplaces of the past.

Other dangers, such as stress, high blood pressure, obesity, and the danger of stroke and heart attack, lurk in the workplace. But tomorrow's headlines may well tell a different story. They may tell of people living longer and enjoying a better quality of life—and work life—of steadily falling health care costs, of increasing percentages of the health care dollar being spent on prevention, and perhaps even of greater long-range corporate profits because employers are wiser and healthier.

These headlines of tomorrow will grow out of the exciting worksite wellness movement today. Everyone who has become deeply involved in the movement develops a sense that history is in the making. Some CEOs report the same sense for their companies: The serious commitment to worksite wellness often seems to coincide with renewed corporate vigor.

Big business will continue to lead the way in research, design, and implementation of worksite wellness programs. The mega-corporations have the resources to do so. They also have millions of dollars to save. Yet small businesses have even more to gain. Their profit margins are usually much smaller. They cannot go for years spending more than they can afford on health care.

As with most other major economic issues, there is much to gain by cooperating. Most large corporations are economically dependent on a larger array of small businesses who provide services and materials. Lee Iacocca was right when he argued for government support for the troubled Chrysler corporation. Iacocca convincingly maintained that not only would Chrysler go under but so would hundreds of small companies. And the thousands of auto workers—whose businesses and jobs were built around the giant corporation—would be in trouble. When big business invests in wellness and shares its knowledge and experience with small business, big business makes an indirect but significant impact on everyone's bottom line.

The place where this kind of partnership between big and small business can be forged is in a Wellness Council. If the councils serve only as clearinghouses of information between businesses about what works and what doesn't work, such coalitions are well worth the investment of time, personnel, and money. But, as has been demonstrated in Omaha and the other cities where councils exist, Wellness Councils benefit the entire community. It started out as just a slogan but turned into a mission. Omahans, especially those in the business sector, mean what they say when they refer to Omaha as the "wellness capital of the world."

Worksite wellness will continue to occupy center stage in the broader wellness movement. The reasons go beyond bottom-line considerations. They even go beyond the obvious impact that such worksite activity has on employee morale. The worksite is the logical place to reach the most people. Also, work is the place where most parents can be found, even parents of young children. Nothing will have an impact on the health of our future generations like good role models provided by healthier parents. But the worksite is the logical place to promote wellness for another, even more compelling reason.

Most of us spend more time with the people at work than we do with our own family, friends, and loved ones. Some social scientists maintain that in this mobile, working society the people at work now fill the needs and offer much of the support that was once provided by large extended families. It has been stressed that wellness is an individual responsibility and must remain so; however, when people are confronted with accurate information and determine that they must change their lifestyles, they need a family-type support group—a climate that encourages wellness. The support that enables people to make life-changing, life-saving decisions can be created most effectively and most efficiently in a healthy company.

With worksite wellness, we have the most compelling opportunity to change the health of the nation.

Wellness may be thought of as a trend today, but it is quickly becoming a fixed pillar of American culture and especially of the American workplace. It promises to have more staying power than rock and roll. Evidence is in the flood of new products, insurance programs, exercise equipment and apparel, health clubs, and even travel opportunities all centered on the wellness consciousness of the American people. Like computer technology—friend or foe—concern for health is here to stay.

People are living longer—that's also good news. Why? The scientist, the doctor, the social scientist, and the insurance executive all agree. The reason for this dramatic gain is that people across the country are generally paying closer attention to their health and well-being. We are not only eating less, but generally we are eating smarter. We are not only exercising more, but we are learning about what kinds of exercise are good for us as individuals. Men particularly are smoking less.

We are also seeking medical help when we need it, instead of "waiting to push the 911 button when we want the physician to take care of what ails us," observes Dr. Donald Darst, a member of WELCOM's medical advisory committee (personal letter to the author).

Medical technology has also made its contribution. For example, the development of safe drugs that can be used to control high blood pressure is believed to be a primary cause of the dramatic drop in fatal strokes.

We Americans in our twenties want to look good, in our thirties we want to feel good, and after 40 we want to live longer. So wellness appeals to consumers of every age. And it appeals not only to the wellness fanatic but to those educated believers who are striving to be more healthy.

As consumers become more knowledgeable and sophisticated, they also learn and really understand that wellness is more than looking good and living longer. It is also about enjoying life more. It is about positive practices that enhance the quality of life. It is about avoiding unnecessary disease and at least postponing serious illness, in what researcher James Fries calls compression of morbidity (Fries and Crapo, 1981). People will be healthier longer, perhaps until they die, rather than being healthy for a while and then suffering prolonged illness before death.

Wellness is a commitment to the truths we have scientifically uncovered about the causes of disease and the principles of good health. It is the acceptance of personal responsibility for the integration of those truths into our everyday lives. It is the ability to delay gratification and expect results only after hard work, to resist the fads and gimmicks that promise overnight success, and to make a lifetime commitment to our own well-being. It is keeping the spiritual, mental, physical, family, and occupational pieces of our lives in harmony, in a more natural balance.

These are not the things of which fads are made. These are values that make sense to perhaps a large majority of Americans. These values will help shape the future of America. What will that future look like? How will the year 2000 reflect the current wellness trends? What will the workplace look like? What is the future of health care? How is worksite wellness related to other trends? Evidence of its value and staying power is shown in the way worksite wellness coexists with other developing trends. Let's explore them.

THE WORKPLACE OF THE FUTURE

This book has of course focused on health promotion in the workplace. There have been many trends identified in recent years that relate to wellness. In fact, if you wanted to be a successful author over the last decade, the formula that worked was to find an attractive way to package some common-sense principles about human behavior in a way that busy, working people and aspiring executives could understand. *Megatrends*, *In Search of Excellence*, and *The One Minute Manager* are the prime examples.

In each of these books, principles are developed that help indirectly to build the case for worksite wellness. Worksite wellness makes sense in light of the developing new workplace. In fact, in subsequent books, authors of these best-sellers spoke more directly about corporate concern for the health of employees. The trends themselves tell us something about what worksite wellness will look like in the year 2000. Here are some of them.

The Changing Work Force

The work force itself is changing. Most dramatic is the tremendous influx of women into the workplace. In times past, Rosie the Riveter performed a job traditionally held by a man because the country was at war. But Rosie isn't going to give up her job and compliantly go home to bear the baby boom this time. She's in the work force to stay. An earlier chapter credited the increase of women in the workplace with helping to pave the way for the worksite wellness movement. Women will help maintain it.

There is another way that working women will shape the future of worksite wellness. Eighty percent of the women in the workplace by 1990, according to the Bureau of Labor Statistics, will have children. Corporate concern for the quality of family life will emerge as an integral part of wellness. Health promotion will include dependents, and concern has already begun to shift slightly from the problems of the working mother to those of the working parent.

The majority of people in the workplace today are members of the baby boom generation. The baby boomers have pushed our society to its limits from the time they overflowed the maternity wards. Today, they have stretched the capacity of the workplace. They find tremendous competition for promotions at the same time that the computer and the world economy are making middle managers obsolete.

The time commitment to work is predicted to change as more and more people flood the labor market for fewer and fewer jobs. A 32-hour work week is envisioned by the year 2000. Finding the time for exercise during the workday will not be so difficult for employee or employer in the workplace of tomorrow.

Across the country, firms are experimenting with flex time and shorter work weeks, with more permanent part-time employment—often of skilled or professional people, with job sharing, with employees working out of their homes, and with flexible benefits. The opportunities to promote worksite wellness will be expanded by the new approaches to time.

Employee assistance programs (EAPs) will become even more prevalent and their mission will broaden beyond just alcohol- and drug-related problems. They will address all types of personal and family problems. Programs that address the concerns of families through worksite wellness and prevention programs are simply the prevention side of EAPs. It stands to reason that less costly prevention is just common sense.

The baby boomers also bring many of their dominant values to the workplace. They continue to have a concern for humanity and justice and want an opportunity to make a meaningful contribution to society. They also express their independence by insisting on the opportunity to have meaningful and satisfying work. Worksite wellness, as described in this book and as it continues to develop in the offices and factories of America, offers a meaningful outlet for many of the unique concerns of the baby boomers. For many companies the wellness program has evolved into a recruiting tool. Perhaps one day a candidate for an executive position will ask not about salary but about the running track, the wellness programs, and the locker rooms.

As they grow older—and live longer—the baby boomers will nudge worksite wellness into helping people plan for and enjoy what will potentially be the most active and purposeful retirement community in the history of the world.

Minorities will make up a larger part of the work force of the future. Blacks will experience upward mobility and more equal access to the world of meaningful employment itself. Immigration will continue to rise, and Hispanics will be the fastest-growing portion of the work force. How is this related to wellness? The benefits of the emphasis on worksite wellness will not be lost on minority employees. In fact, minorities suffer disproportionately in every major health indicator— from mortality rates to high blood pressure to accidental deaths and problems related to drug and alcohol abuse. More minority people will be found in the work force than in any other single place in our society. What better place to reach them and educate them about health and wellness?

Information, Service, and High Technology

Wellness is here to stay because it is developing in the most educated and information-absorbing society in history. The high-tech environment creates unique kinds of stress. It requires that people spend large amounts of time in front of computer terminals. Not only is the work sedentary, but it creates numerous physical strains. Concern has arisen about the long-term effects on eyesight. Anyone who has spent much time in front of a terminal gains new insight into the phrases "pain in the neck" and "my aching back." Medical people and high-tech manufacturers will seek to improve the equipment itself, to try to ensure that it is easy on the eyes. Experimentation is already taking place with new kinds of chairs that help to prevent neck and back strain. These things will help to create the ergonomic (adapting work to the worker) office of the future, but a part of the solution will also be the emphasis on wellness to promote, not endanger, health. Employees and supervisors have already discovered the positive effects of an exercise break, relaxation techniques, and stretching for people who spend most of their time interacting with a computer terminal.

The nature of high tech itself creates a continuum of stress, and wellness may be the best answer at each extreme of the continuum. On one end are the highly sophisticated tasks of systems planning and programming. The highly trained people who perform these tasks are required to solve difficult problems that few untrained people can even begin to comprehend. Usually they are working under demanding deadlines driven by fierce competition. At the other end of the spectrum are unskilled people who are hired to perform simple mundane tasks, with no real understanding of the computer at all. They are often required to do the same thing—enter information from forms, verify credit card numbers over the phone—day in and day out, year in and year out. They are often coldly measured by their speed and accuracy. These modern automatons are not so far removed from the humdrum of the assembly line—putting the same nuts and bolts into the same piece of equipment, time after never-ending time. Worksite wellness not only provides the necessary breaks from the pressure, it also communicates that the company cares about the employee and that people are more important than machines.

The New Professionals

By 2000 wellness and health promotion will employ large numbers of people. According to labor statistics, occupational therapists, health sciences administrators, physical therapists, podiatrists, and medical assistants will experience the highest growth rates of all occupations. Their ranks will be swelled by the "new professionals" or those people using their education and skills for preventive medicine and health promotion.

Sometimes referred to as "alternative practitioners," their emergence on the scene has caused a dramatic change in undergraduate and graduate programs all across the country. The traditional health, physical education, and recreation (HPER) departments are quickly changing, rearranging, and adding to their curricula in order to respond to the growing demand for wellness practitioners. Career opportunities for HPER majors used to be limited to teaching or nonprofit agency work. Today opportunities are much broader and promise to grow even more. Corporate and private fitness centers, insurance companies, financial service institutions, governmental agencies, and HMOs are all looking more and more for trained wellness practitioners to lead their educational and prevention efforts. Traditional agencies like the YMCA are also redesigning their programs to be able to adequately respond to the new interests that go beyond fitness.

Springfield College in Springfield, Mass., was founded in 1885 as a school for YMCA workers. It is respected for its leadership over the years in physical education and recreation. In 1985 it graduated 100 students with degrees in health-fitness. From 50 students in 1980, the program grew to 450 students by 1985. Many graduates still go to work for the Y and other nonprofit agencies. But other students have found jobs at Prudential Life Insurance, Xerox, and private health clubs.

"Exercise science and health promotion is the fastest growing track within our profession," says Kris Berg, chairman of the HPER department at the University of Nebraska at Omaha and a founding member of WELCOM. "Organizations such as WELCOM are at the forefront of changing the health profile of the country, and our profession is closely watching its activities. Our students hope to play a vital role in implementing worksite wellness programs" (personal letter to the author).

Culture on Culture

Earlier chapters introduced the SANE approach to a healthy company. It was suggested that companies begin with simple, low-cost programs that educate people about smoking, alcohol, nutrition, and exercise. The whole idea is to educate and encourage employees to make changes. Creating a climate that encourages change is another way of saying the same thing. The climate at a company is sometimes referred to as the corporate culture. Sociology teaches us that the role of a culture is to teach and pass on values from one generation to another. A culture that has a strong climate with clearly defined values and practices that are consistent with those values is a powerful influence on individual and group behavior. In essence that is what worksite wellness is—creating a strong culture that influences individuals and groups to take personal responsibility for their own well-being.

There is another side of the same coin. Every corporation or business operates within a much larger culture, sometimes referred to as the American way of life. The larger culture also influences the smaller culture in many dramatic ways. There are trends taking place within the larger societal context which are also having an impact on corporate America. Like it or not, the workplace will have to adjust. Sometimes it is difficult to tell which culture is influencing which. It is usually some of both.

For example, Americans generally are changing their eating habits and will continue to do so. In some corporations the trend is being encouraged and employees are learning about nutrition in worksite wellness programs. In other worksites, management that has not recognized the trend will find itself under pressure from employee groups to make the cafeteria food more healthful. The pressure may be direct and demanding, perhaps through an official grievance process. Or it may be subtle. Employees may simply begin to eat out more often. Therefore, companies that today do not recognize the importance of worksite wellness may tomorrow find themselves under pressure from employees and unions to respond to those very same issues.

One issue in particular is drug abuse and employer-sponsored drug testing. Corporations will have to respond to this issue long before the year 2000. The possibility of creating a drug-free workplace should be addressed positively and with a cooperative spirit by employers and employees.

DRAMATIC SAVINGS: IN LIVES OR MONEY?

Headlines will continue to showcase technical advances like the artificial heart, while at the same time spotlighting the dramatic savings brought about by good preventive medicine. Although advances will continue to be made in both arenas, experts predict that, before long, we will be forced to choose between medical miracles and saving money.

Dr. Pierre Galleti of Brown University believes there is no organ which won't be replaced in the future. He and others believe that by the year 2000, not only will many people live with miniaturized artificial hearts, but many others will have artificial lungs, kidneys, pancreases, blood vessels, ears and maybe even eyes (*Omaha World-Herald*, November 1985).

The potential for better health as the result of advances in medical technology is difficult to comprehend. Research will continue and probably accelerate. Few people are aware of the volume or the scope of the research taking place. In addition to the promises held forth by artificial body parts, there is tremendous potential for good in the explosion of research in genetics and neurology.

But as exciting as it all is, the more we learn, the more we come to understand two important things. First, there are limits. Few diseases have been researched with such commitment as cancer. Advances have been made, and future gains are expected, but the disease will not disappear. Most of the advances that have been made have resulted more from early detection and prevention, rather than from treatment and cure. The National Cancer Institute foresees that by 2000 some cancers like breast and ovarian will be highly curable. Lung cancer, however, will remain a major killer. And a troubling cloud that hangs over the future is the AIDS virus, which can lead to some types of cancer. The ravaging effects of cancer and some other diseases will remain with us well into the next century and beyond.

The second thing we know is that we will not be able to afford all of the research and technology. With the total health care bill climbing beyond $400 billion a year, and only a tiny fraction of that amount being spent on the prevention of disease, the question of ethics must again be raised. Is it fair? Should millions be spent on risky technology and dramatic procedures that for decades to come hold promise for only a few lucky ones, when much less money, wisely spent in educational and prevention programs, can benefit thousands almost immediately? As exciting as the technology is, there is a growing conviction that environmental changes, along with good health promotion, lead to behavior changes by large segments of the population. Here is the greatest potential for improving the health of Americans. But the struggle between both sides of this noble spectrum will continue for years to come.

Once again the worksite will play a major role because businesses pay a major share of our total health care costs. Also confronted by the age of limitations, business will speak out loudly of their concerns about health care costs.

Related to this will be continued experimentation with alternate forms of health care insurance. HMOs and PPOs will come and go. They will look different in 2000, but whatever names are given them, health insurance packages will continue to evolve and will increasingly encourage and reward good health. Insurers, hospitals, clinics, and doctors will merge into mega-companies. These conglomerates will offer cut-rate insurance plans that entice companies and encourage patients to stay well. The impetus for this will come from more than the business sector. Nonprofits and government employers may be even more concerned because they have no profits to cover rising health care costs and no way to pass off costs to consumers.

It is ironic that at the same time all of this activity is taking place, a surplus of doctors is predicted to hit the market as early as 1990. In some ways doctors and hospitals appear to be the losers in this health care revolution. With competition heating up and the alphabet health care plans taking over (HMO, PPO, IPP, DRG), doctors may find themselves working for salaries and their waiting rooms filled with patients. As health promotion and disease prevention take their place in the curriculum of the nation's medical schools, more doctors themselves will become role models for a healthier lifestyle. Tomorrow's patients will seek medical professionals who practice a new kind of health care. Patients will look to health care practitioners who themselves practice healthy lifestyles.

TOWARD A SMOKE-FREE SOCIETY

Surgeon General C. Everett Koop, M.D., wants to see a smoke-free society by the year 2000. Dr. Koop said in several interviews following his nomination by President Reagan to a second four-year term that he knows that if his campaign is successful a lot of people will be hurt in the short run. These include the farmer, the distributor, the retailer, and the wholesaler. The American Medical Association agrees that over the long haul, we have no choice but to work toward the goal (*Omaha World-Herald*, October 1985) Why? Is it worth it?

"If AIDS was killing 350,000 Americans a year, efforts to control the plague would be at crisis level," writes columnist Joan Beck in the *Chicago Tribune*. She adds, "If any other product was known to cause 125,000 deaths from cancer every year, it would be off the market before you could say 'cyclamate' or 'EDB.'" Why aren't the same standards applied to cigarettes, she asks, which are the single biggest source of preventable illness and premature death today? What can be done to curb an industry that annually costs the nation $16 billion in extra medical care and $37 billion in lost productivity and earnings because of illness, disability, and death?

Employers again bear much of the costs for smoking. It is not hard to understand why corporate America has begun to support the smoking cessation effort. In fact, what was once considered impossible—restricting or even banning smoking at work—is becoming a prudent thing to do. Many companies are adopting smoking policies. Some policies are simply a directive to supervisors to do what they can to separate the smoking and nonsmoking employees. Other policies restrict smoking to certain areas of the building, usually not allowing smoking at the work station or in conference rooms. A few companies have decided to ban smoking entirely from their premises. And court decisions have even backed the rights of employers not to hire smokers.

A smoke-free society by the year 2000? Probably not. But Surgeon General Koop is on target when he says, "I think the person who smokes in 1995 is going to have to smoke alone or with other smokers. That might be in his bathroom or his backyard or in a segregated area. I wouldn't be surprised to see it out of doors."

Wellness in the year 2000? Wellness, the acceptance of personal responsibility for one's own well-being, represents the best in American culture. Wellness, which not only adds years to our life but life to our years, makes sense. To strive to take charge of our own destinies, to work toward a better quality of life, whether we are rich or poor, young or old, regardless of race, creed, or religion, is something that is touching the soul of the American character.

The best of American business and industry, both large and small, has a historic opportunity to join hands with the best of the American medical community and allied professionals and to help shape what may already be the most significant movement of the last 20 years of the twentieth century: worksite wellness.

APPENDIX A

Bylaws of
The Wellness Council of the Midlands

ARTICLE I MEMBERSHIP

1. Membership in The Wellness Council of the Midlands shall be extended to any employer in the Midlands seeking and being approved for membership.

2. Membership is divided into two classes: Corporate and Business; and Non-Profit/Educational.

3. If the employer has 1,000 or more employees, the membership contribution shall be $2,000. If the employer has more than 125 but less than 1,000 employees, the membership contribution shall be $2 per employee. If the employer has fewer than 125 employees, the membership contribution shall be $250.

4. The membership contribution for Non-Profit/Education class institutions shall be $250.

5. All members, regardless of membership class or contribution, shall have equal rights as set forth in these bylaws.

6. Members must pay annual dues as assessed by the Board of Directors to maintain member status. If a member's dues are found to be in arrears for a period of ninety (90) days or longer, the Board of Directors may take whatever action it deems necessary or appropriate to rectify the delinquent payment.

7. Each corporate member shall designate in writing the person it desires to represent it at the member meetings. A representative and, if desired by the member, an alternate representative may be appointed by a member. A corporate member may replace a representative, or an alternate representative, by written notice to the President and designate a new representative or a new alternate representative.

ARTICLE II MEMBERS' MEETINGS

1. An annual meeting of the members shall be held either within or without the State of Nebraska as may be provided in the Notice of Annual Meeting to members.

2. The annual meeting of the members shall be the first members' meeting of each calendar year, but in no event shall the annual meeting occur later than March 31st of each year.

3. Failure to hold the annual meeting of the members at the designated time shall not work a forfeiture or dissolution of the Council.

4. Special meetings of members may be called by the President or by the Board of Directors or by members having one-half of the votes entitled to be cast at such meeting.

5. Notice stating the place, day and hour of the meeting and, in the case of a special meeting, the purpose or purposes for which the meeting is called, shall be delivered not less than 10 nor more than 50 days before the date of the meeting, either personally or by mail, by or at the direction of the President, or the Secretary, or persons calling the meeting, to each member entitled to vote at such meeting. If mailed, such notice may be included within any regular publication mailed to the members and shall be deemed to be delivered when deposited in the United States mail addressed to the member at his address as it appears on the records of the Council, with postage prepaid.

6. No notice need be given of regular meetings or of adjourned meetings, other than the annual meeting of members.

7. Representation in person or by proxy of one-tenth of the members at the annual or a special meeting shall constitute a quorum.

ARTICLE III MEMBERS' VOTING

1. The vote of a majority of members present or represented by proxy at a meeting at which a quorum is present shall be necessary for the adoption of any matter voted upon by the members. Each member shall have one vote.

2. A member may vote in person or by proxy executed in writing by an officer of the member or its duly-authorized representative. No proxy shall be valid after eleven months from the date of its execution unless otherwise provided in the proxy.

3. Members are entitled to vote with respect to the election of the Board of Directors and upon any such other matter as may be submitted to the membership for a vote at the annual meeting or a special meeting as provided in the notice of such special meeting.

ARTICLE IV   BOARD OF DIRECTORS

1. The direction and management of the affairs of The Wellness Council shall be vested in a Board of Directors. The Board of Directors shall exercise all the corporate powers of The Wellness Council as may be necessary to carry out its purposes.

2. The number of directors shall not be less than three nor more than twelve.

3. The directors need not be residents of the State of Nebraska. No member shall have more than one representative on the Board.

4. Directors shall be elected by the members at the annual members meeting.

5. Each director shall hold office for the term for which he or she is elected and until his successor shall have been elected and qualified. Any vacancy occurring in the Board of Directors may be filled by a majority vote of the Board of Directors to fill the vacancy for the unexpired term of any director.

6. The first Board of Directors elected by the members (as distinguished by the Board of Directors appointed in the Articles of Incorporation) shall consist of twelve members. Four persons shall be elected to serve as directors for a three-year term of office. Four persons shall be elected to serve as directors for a two-year term of office. Four persons shall be elected to serve as directors for a one-year term of office.

7. Subsequent to the first member-elected Board of Directors, each successive year four persons shall be elected to the Board of Directors for a three-year term.

8. All directors shall be eligible for re-election as directors.

9. A majority of the number of directors having been elected at any time shall constitute a quorum for the transaction of business.

10. The initial officers of the corporation shall be selected from the members of the Board of Directors at the time of the election of officers.

11. The directors appointed as such in the Articles of Incorporation shall serve a one-year term unless replaced or substituted prior to the expiration of such term by the members at a special meeting of the members called for this purpose.

ARTICLE V   MEETINGS OF THE BOARD OF DIRECTORS

1. An annual meeting of the Board of Directors shall be held immediately following the annual meeting of the members at the location of the membership meeting.

2. Additional meetings of the Board of Directors may be held at any time upon the call of the President or any three directors, by written notice mailed to each director at least seven days prior to the meeting. Notice of meetings may be waived in writing and shall be considered to be waived by all directors at any meeting at which all directors are present.

3. At any meeting of the Board of Directors, a majority of the entire Board shall constitute a quorum for the transaction of business. The act of a majority of the directors present shall be the act of the Board of Directors except as otherwise set forth herein. "Entire Board" shall mean the number of directors the Council would have in the absence of a vacancy.

ARTICLE VI   OFFICERS

1. The officers of the corporation shall consist of a Chairman of the Board, a President, one or more Vice-Presidents, Secretary, Treasurer, and such other officers and assistant officers as may be deemed necessary by the Board of Directors. Any two or more offices may be held by the same person, except the offices of President and Secretary. The officers shall be elected at an annual meeting of the Board and shall serve for a term of two years or until their successors are elected and qualified. Except for the President, and except for the first officers elected by the Board, an officer need not necessarily be a director. Any officer may be removed from office with or without cause by the Board of Directors at any time.

2. The President shall execute on behalf of the corporation such contracts and other documents as may be authorized by the Board of Directors and shall be responsible for day-to-day activities of the corporation.

3. The Chairman of the Board shall preside at all meetings of the Board of Directors. The Chairman is the chief spokesman and director of the corporation, and it is his direct responsibility to coordinate the work of all committees, the actions of the Board, the work of the officers, and the work of the Executive Director, as well as to have overall responsibility for directing membership meetings and membership events. It is contemplated that these responsibilities of the Chairman shall be fulfilled by him through delegation of duties to the President, committee chairpersons, other officers of the corporation, and the Executive Director.

4. If a vacancy occurs in the office of the President for any cause, or if the President be absent or unable to perform the duties of the office, the Vice-President shall have the power to perform the duties of the President.

5. The Treasurer shall have charge of and safely keep valuable papers and securities of the corporations, shall collect and deposit all monies in such depositories as may be designated by the Board and, with such additional signatures as the Board may prescribe, disburse the funds of the corporation as directed by the Board of Directors. The Treasurer shall also have charge of the books of account, and enter or cause to be entered therein a true statement of all receipts and disbursements. The Treasurer shall make a report of the affairs of the corporation within the Treasurer's charge at the annual meeting of the directors, the annual meeting of the membership and at such other times as the Board of Directors may require.

6. The Secretary shall keep, or cause to be kept, a record of the proceedings at meetings of the Board of Directors and shall promptly give all notices of directors' meetings in accordance with the bylaws. The Secretary shall be the custodian of the corporate seal and affix same to and attest to all instruments executed by the corporation which require attestation and the affixing of the corporate seal. The Secretary shall attend to the publication of all notices and shall make a report of the affairs of the corporation within his charge at the annual meeting of the directors, membership and at such other times as the directors may require.

ARTICLE VII   COMMITTEES

1. The Board of Directors may provide for such standing and other committees as it shall deem wise and the Board of Directors may delegate to such committees such duties and powers from time to time as it shall deem necessary or desirable.

ARTICLE VIII   EXECUTIVE DIRECTOR

1. The Board of Directors shall have authority to select an Executive Director who shall be the chief administrative officer of the corporation. The Executive Director shall be paid such salary as the directors shall determine and shall be responsible for the execution of such plans and policies as the Board of Directors may authorize, direct or approve. The Executive Director shall appoint members of the administrative staff, if any, and assign them appropriate duties. This person shall serve ex-officio, but without vote, on the Board of Directors and all committees appointed by the Board of Directors.

ARTICLE IX   FISCAL YEAR

1. The fiscal year of the corporation shall be the calendar year unless the Board of Directors shall otherwise determine.

ARTICLE X   AMENDMENTS

1. The bylaws may be altered or amended by the Board of Directors provided that no such amendment shall be inconsistent with the Articles of Incorporation and no such amendment shall in any way change the non-profit character of the corporation. Notice in writing of a proposed amendment shall be given to said directors at least seven days prior to the meeting of the Board of Directors at which such amendment is proposed to be adopted, unless all members of the Board eligible to vote on such amendment unanimously agree in writing to adopt such amendment, or unless all such members of the Board are present at a meeting at which such amendment is adopted by unanimous vote.

Articles of Incorporation of
The Wellness Council of the Midlands

We, the undersigned, desiring to form a corporation under the Nebraska Nonprofit Corporation Act do hereby adopt the following Articles of Incorporation.

ARTICLE I

The Name of the corporation is The Wellness Council of the Midlands.

ARTICLE II

The corporation shall commence business on the date of filing of the Articles of Incorporation and shall have perpetual existence.

ARTICLE III

The corporation is organized for the benefit of all employees of businesses and industries, as well as, any other civic or nonprofit entities, in the Midlands. The corporation shall be under the control of a Board of Directors represented by member companies. The purposes and objectives for which the corporation is organized are as follows:

1. To foster the wellness disciplines throughout corporate communities operating in the Midlands in order that medically sound fitness information and action programs designed to alter unhealthy lifestyles may become more and more available in the workplace.

2. To promote cooperative arrangements within the corporate community and to pool resources and information aiding the advancement of effective wellness programs.

3. To promote educational programs for top corporate management regarding the advantages and pay-offs inherent in wellness disciplines.

4. To seek cooperative agreements with other organizations in the health field for the purpose of advancing the wellness disciplines.

5. To provide a clearinghouse of information regarding the wellness resources available as well as historical information regarding experiences of other entities involved in wellness activities.

ARTICLE IV

The corporation shall have the specific power to hold property in trust for itself and for the carrying out of any of its authorized purposes and shall have the powers which now are or hereafter may be conferred by the Nonprofit Corporation Act of Nebraska, subject to the following limitations and conditions which shall apply notwithstanding any other provisions of these articles. To wit: no part of the net earnings of the corporation shall inure to the benefit of or be distributable to its members, directors, officers or other private persons, except that the corporation shall be authorized and empowered to pay reasonable compensation for services rendered and to make payments and distributions in furtherance of the purposes set forth in Article III hereof: no substantial part of the activities of the corporation shall be the carrying on of lobbying, or otherwise attempting, to influence legislation, and the corporation shall not participate in, or intervene in (including the publishing or distribution of statements) any political campaign on behalf of any candidate for public office. Notwithstanding any other provision of these Articles, the corporation shall not carry on any other activities not permitted to be carried on (A) by a corporation exempt from federal income tax under section _____ of the Internal Revenue Code of _____ (or the corresponding provision of any future United States Internal Revenue Law) or (B) by a corporation, contributions to which are deductible under section _____ of the Internal Revenue Code of _____ (or the corresponding provision of any future United States Internal law).

Upon the dissolution of the corporation, the Board of Directors shall, after paying or making provisions to pay all of the liabilities of the corporation to educational, scientific, or charitable organizations which are then exempt from federal income tax under the provisions of section _____ of the Internal Revenue Code of _____ or any successor provisions, as the Board of Directors shall determine.

ARTICLE V

The initial registered office of the corporation is _____.

The initial registered agent of the corporation at such address is _____.

ARTICLE VI

The corporation shall have members, its affairs shall be managed by a Board of Directors whose number and manner of election or appointment shall be provided in the bylaws of the corporation, but in no case shall the number be less than three. The initial members of the Board of Directors shall be:

ARTICLE VII

The name and address of each incorporator is:

ARTICLE VIII

These Articles of Incorporation may be amended from time to time by the Board of Directors in the manner provided for by the Nebraska Nonprofit Corporation Act, provided that under no circumstances shall any amendment change the nonprofit and the educational, scientific and charitable character of the corporation.

In witness whereof, the incorporators have signed and adopted these Articles of Incorporation on this _____ day of _____, 19____.

APPENDIX B

JOB DESCRIPTION: EXECUTIVE DIRECTOR OF THE WELLNESS COUNCIL

TITLE: Executive Director

The Executive Director is responsible for the administrative functions of the Wellness Council.

DUTIES:

Establish and maintain working relationship with council delegates.
Oversee and advise committees.
Manage Wellness Council office.
Supervise Assistant Director and other personnel.
Direct budget operations.
Organize delegate meetings.
Oversee council publications.
Develop and implement marketing strategy.
Provide for the dissemination of general information.
Administer membership recruitment activity.

SKILLS:

Public speaking
Interpersonal communications
Organization

JOB DESCRIPTION: ASSISTANT DIRECTOR OF THE WELLNESS COUNCIL

TITLE: Assistant Director

The Assistant Director will work closely with and under the direct supervision of the Executive Director.

DUTIES:

Acts as editor of the newsletter and other publications.

Serves as an ex-officio member of the special activities committee and the communications committee.

Attends various other committee meetings as requested by the Executive Director.

Responsible for billing and keeping balance sheets.

Sends out notices, handles registration, assists in menu planning, and confirms sites for board meetings and delegate meetings.

Sends out notices of other committee meetings.

Maintains an up-to-date delegate list.

Initiates and answers correspondence.

Performs routine office procedures (filing, typing, mail).

Completes any other various projects or tasks requested by Executive Director.

REFERENCES

Is American Business Ready to Make a Difference? Remarks from Ronald Reagan, President of the United States, during a video teleconference, "Wellness and the Bottom Line," March 1984. Used with permission.

Preface

Quote by T George Harris. In "Fitness, Corporate Style," *Newsweek*, November 5, 1984. Used with permission.

Introduction

Quote by Roger B. Smith, Chairman, General Motors Corp., at conference on *Worksite Health Promotion and Human Resources: A Hard Look at the Data.* Washington, D.C., October 1983. Used with permission.

Quote by John Pinney, "Corporate Smoking Policies Clearing the Air," *Response*, May 1985, p. 19. Used with permission.

Chapter 1

Parts of this chapter were adapted and used with permission from Creighton University *Window* magazine.

Quote from *Re-inventing the Corporation.* Copyright 1985 by John Naisbitt and Patricia Aburdene. Published by Warner Books.

Quote by Joseph A. Califano, Jr., from *Chrysler's Health Care Cost Containment: The First Results.* Remarks before the Health Insurance Association of America, Kansas City, April 1985.

Quote by Robert N. Beck, "The Name of the Game Is Change," *Esprit.* Winter 1985/1986, p. 5. Copyright 1986 Central States Health & Life Co of Omaha.

Quote by Regina E. Herzlinger. Reprinted by permission of the *Harvard Business Review.* Excerpts from "How companies tackle health care costs: Part II" by Regina E. Herzlinger (September-October 1985). Copyright 1985 by the President and Fellows of Harvard College; all rights reserved.

"The Role of Corporate America" is adapted from *Families: We've Never Been Here Before.* Copyright 1985 by Fred W. Schott. Published by Central States Health & Life Co. of Omaha.

Chapter 2

Quote from Blanchard, M., and Tager, M. *Working Well.* New York: Simon and Schuster, 1985. Copyright 1985 by Marjorie Blanchard and Mark Tager.

Healthstyle is a government publication used with permission from the National Health Information Clearinghouse.

The Health Risk Appraisal is used with permission and is available from the Centers for Disease Control in Atlanta.

Chapter 5

Excerpts by Charles A. Berry, M.D. Used with permission.

Chapter 6

Quote by Arthur L. Ulene, M.D., from HIAA conference *Health Education and Promotion: Agenda for the 1980s,* Atlanta, 1980. Used with permission.

Quote by James R. Brennan. Statement of the Health Insurance Association of America. "Health Promotion Initiatives of the Health Insurance Industry" presented before the Subcommittee on Health, Senate Finance Committee, U.S. Senate, June 14, 1985. Used with permission.

Quote by Lynn H. Johnston. "Good Reason to Get Excited about Wellness, *National Underwriter,* December 14, 1985, pp. 11, 19.

Chapter 7

Quote by columnist Joan Beck in "Cigarette Producers Should Be Target." Copyright 1985 Chicago Tribune Company, all rights reserved, used with permission.

SOURCES CITED

American Cancer Society. "The Economic Impact of Cancer and Cancer Control on Private Industry," 1981.

Belloc, N. B. and L. Breslow. "Relationship of physical health status and health practices." *Preventive Medicine, 1,* 409–421 (1972).

Berry, Charles A. and Michael A. Berry. "Wellness: A Positive Strategy for a Healthy Business." *Time,* June 18, 1984 (special advertising section, "Rx for Controlling Employee Health Care Costs").

Blanchard, Kenneth and Spencer Johnson. *The One Minute Manager.* New York: Morrow, 1982.

Blanchard, Marjorie and Mark Tager. *Working Well.* New York: Simon and Schuster, 1985, p. 41.

Chamber of Commerce of the United States. *Directory of Business Coalitions for Health Action.* Washington, D.C.: Chamber of Commerce of the United States, 1983.

Corporate Fitness & Recreation. "Corporations Help Employees Quit," June/July 1985, p. 9.

Eliot, Robert S. and Dennis L. Breo. *Is It Worth Dying For? A Self-Assessment Program to Make Stress Work for You Not against You.* New York: Bantam, 1984.

Fries, J. F. and L. M. Crapo. *Vitality and Aging.* San Francisco: W. H. Freeman, 1981.

HealthWorks Northwest. *Health Promotion Programs in Small Businesses.* Seattle, Wash.: Puget Sound Health Systems Agency, 1984.

Herzlinger, Regina E. "How Companies Tackle Health Care Costs: Part II." *Harvard Business Review,* September-October 1985, pp. 108–120.

Herzlinger, Regina E. and David Calkins. "How Companies Tackle Health Care Costs: Part III." *Harvard Business Review*, January-February 1986, pp. 70–80.

Herzlinger, Regina E. and Jeffrey Schwartz. "How Companies Tackle Health Care Costs: Part I." *Harvard Business Review*, July-August 1985, pp. 68–81.

Hewitt Associates. *Company Practices in Health Care Cost Management*. Lincolnshire, Ill.: Hewitt Associates, 1985.

Kristein, Marvin M. "How Much Can Business Expect to Profit from Smoking Cessation?" *Preventive Medicine, 12*, 358–381 (1983).

Larson, Arthur. *The Law of Workmen's Compensation*. Albany, N.Y.: Matthew Bender, 1972.

Maccoby, Michael. *The Leader*. New York: Simon and Schuster, 1981.

Naisbitt, John. *Megatrends*. New York: Warner Books, 1982.

Naisbitt, John and Patricia Aburdene. *Re-inventing the Corporation*. New York: Warner Books, 1985.

Newsweek. "Fitness, Corporate Style," November 5, 1984, pp. 96–97.

New York State Journal of Medicine. "Confronting America's Most Costly Health Problem, A Dialogue with Surgeon General Koop," December 1983, pp. 1260–1263.

Omaha World-Herald. "Smoking Burns Holes in Economy," September 17, 1985.

Omaha World-Herald. "Report: Minorities Less Healthy, but Money Not Solution," October 17, 1985, p. 7.

Omaha World-Herald. "Peer Pressure to Snuff Out Smoking by 2000," October 21, 1985, p. 1.

Omaha World-Herald. "Replacing Human Body Parts to Be Even More Widespread," November 17, 1985, p. R1.

Ouchi, William G. *Theory Z: How American Business Can Meet the Japanese Challenge*. Reading, Mass.: Addison-Wesley, 1981.

Peters, Thomas J. and Robert H. Waterman, Jr. *In Search of Excellence: Lessons from America's Best-Run Companies*. New York: Warner Books, 1982.

Peters, Thomas J. and Nancy Austin. *A Passion for Excellence: The Leadership Difference*. New York: Random House, 1985.

Promoting Health/Preventing Disease: Objectives for the Nation. Public Health Service, Department of Health and Human Services. Washington, D.C.: U.S. Government Printing Office, 1980.

Townsend, Robert. *Further Up the Organization*. New York: Knopf, 1984.

U.S. News & World Report. "Tomorrow," October 14, 1985, p. 18.

Wall Street Journal. "On the Run," August 19, 1985, p. 21.

RECOMMENDED READING

Much information is produced by various government and health-affiliated agencies and tends to become dated. Therefore, these items will not be included in this reference section. Most of these organizations are in the Washington, D.C., area. Some of those who regularly produce health information that may include worksite wellness are the Institute of Medicine (National Academy of Sciences), President's Council on Physical Fitness and Sports, Office of Disease Prevention and Health Promotion (DHHS), and several clearinghouses such as

National Health Information Clearinghouse
P.O. Box 1133
Washington, DC 20013-1133
phone: 1-800-336-4797

Clearinghouse on Business Coalitions for Health Action
U.S. Chamber of Commerce
1615 H St., N.W.
Washington, DC 20062

In addition, several professional and consumer periodicals carry health information articles that are relevant to wellness at the worksite. Some of the magazines that carry such articles are *U.S. News, Business Week,* and *Time*. Specific magazines and newsletters that deal with issues of worksite wellness include *American Health, Optimal Health, Corporate Fitness and Recreation, Business*

Special thanks to Karen Hackleman, consultation coordinator for the Southeastern/Atlantic Regional Medical Library Services in Baltimore, and Kelly Jennings, health information librarian for the Tulsa City-County Library System, for selecting the best and most-asked-for books and writing the descriptions of them for this section.

169

and Health, Executive Fitness Newsletter, Wellness Management, Employee Health & Fitness, Corporate Commentary, and *Prevention,* among many others.

CHAPTER 1: FITNESS THRIVES FROM 8 to 5

Berry, Charles A. *An Approach to Good Health for Employees and Reduced Health Care Costs for Industry.* Houston: National Foundation for the Prevention of Disease. $5

Frank talk from a medical doctor and specialist in preventive medicine to business leaders about risk factors and health promotion strategies. Dr. Berry is the former medical director of the U.S. space program. Order this booklet from the National Foundation for the Prevention of Disease, 10777 Westheimer Road, Suite 935, Houston, TX 77042.

Blanchard, Marjorie and Mark J. Tager. *Working Well: Managing for Health and High Performance.* New York: Simon and Schuster, 1985. $15.95

The authors' research indicates that bad management can create unhealthy employees. Offers simple managerial tactics that any manager can use to ensure a healthy, motivated work force. Introduces the PERKS model: participation, environment, recognition, knowledge, and style. Foreword by Kenneth Blanchard, co-author of *The One-Minute Manager.*

Brennan, Andrew (editor). *Worksite Health Promotion.* New York: Human Sciences Press, 1982. $9.95

Special issue of *Health Education Quarterly,* a journal, devoted to discussing education for improving the personal health behavior of employees through programs provided by the employer.

Cunningham, Robert M., Jr. *Wellness at Work.* Chicago: Blue Cross/Blue Shield, 1982. $7.95

Report on health and fitness programs for employees of business and industry. Cites yearly costs and Blue Cross/Blue Shield public health education programs. Case studies at Xerox, IBM, Control Data, Campbell's Soup, and Bonne Bell are reported. Author has written other health-related books.

Everly, George S., Jr. and Robert H. L. Feldman and Associates. *Occupational Health Promotion: Health Behavior in the Workplace.* New York: Wiley, 1985. $27.95

Guide for practitioners and administrators working in corporate, business, and labor settings. Includes factors that affect health compliance in the workplace and worker satisfaction. Appendix describes health programs located at Control Data (Staywell), IBM, and Johnson & Johnson (Live for Life).

Fielding, Jonathan E. *Corporate Health Management*. New York: Addison-Wesley, 1984.

A technical but complete book for personnel/benefits people who want to look at the insurance options available to revamp current health insurance plans. A financial discussion, sometimes very detailed, but complete and helpful. The best for this purpose.

Matarazzo, Joseph D., Sharlene M. Weiss, J. Alan Herd, Neal E. Miller, and Stephen M. Weiss (editors). *Behavioral Health: A Handbook of Health Enhancement and Disease Prevention*. New York: Wiley, 1984.

Academic text discusses behavioral health models: exercise, diet, smoking prevention, blood pressure, dental health, and others, and the settings for health promotion including the worksite. Reviews training for health promotion programs and outlook for the next century, in 95 chapters.

O'Donnell, Michael P. and Thomas H. Ainsworth (editors). *Health Promotion in the Workplace*. New York: Wiley, 1984. $36

Discusses all the major components and critical issues involved in workplace health promotion. Included are interviews with CEOs of companies that currently have health promotion programs. Also outlines systematic methods for selling the concept of wellness to employers and upper management. Good reference tool, highly detailed.

Opatz, Joseph P. *A Primer of Health Promotion: Creating Healthy Organizational Cultures*. Washington, D.C.: Oryn Publications, 1985. $21.95

Calls for the sharing of resources between corporations and community organizations to enhance the efforts of the individual organization. A short scholarly discussion, not a practical guide.

Yenney, Sharon L. *Small Businesses and Health Promotion: The Prospects Look Good*. New York: National Center for Health Education, 1984.

Forty-six-page guide for providers of health promotion programs. Especially good advice for small business owners on how to start with easy programs and how to collaborate with other small businesses. Study conducted through the Office of Disease Prevention and Health Promotion.

CHAPTER 2: A BUSINESS PLAN FOR HEALTH PROMOTION AT WORK

HealthWorks Northwest, Puget Sound Health Systems Agency, 601 Valley St., Seattle, WA 98109, $15 each

Four publications that include helpful checklists and questionnaires. No doubt the best working guidebooks available for nuts-and-bolts information.

> *Employee Health Promotion: A Guide for Starting Programs at the Workplace* (1983): A step-by-step approach with worksheets and decision points. Includes graphics on health promotion for use in overhead projectors.
>
> *Health Promotion Needs Assessment Manual* (1984): Designed for employers and others involved in getting worksite programs started. Specific interest in the interview series—questionnaires for the CEO, directors of personnel and benefits, management, and employees.
>
> *Health Promotion Programs in Small Businesses* (1984): Information on the development and variety of health promotion programs in small businesses throughout the country. Survey results on the benefits of health promotion for small businesses. Several informative case studies.
>
> *Guidelines for Selecting Health Promotion Providers* (1984): Subject-specific guidelines and sample questions to ask when seeking a vendor for programs in smoking cessation, alcohol and drug abuse, nutrition, stress management, and other program areas.

Parkinson, Rebecca S. and Associates. *Managing Health Promotion in the Workplace: Guidelines for Implementation and Evaluation.* Palo Alto, Calif.: Mayfield, 1982. $24.95

Developed by a group of experts in worksite health promotion, this book provides a series of guidelines designed to help individuals plan, implement, and evaluate effective employee health promotion programs. It also addresses such management issues as where the program should be located within the organizational structure and how to identify available resources. In addition, it contains examples of programs in 17 corporations, plus 12 background papers on topics including hypertension, weight and nutrition, alcohol and drug abuse, smoking cessation, stress management, physical activity, and cost-effectiveness.

CHAPTER 3: THE SANE APPROACH TO A HEALTHY COMPANY

Executive Health Examiners. Richard E. Winter (editor). New York: McGraw-Hill, 1983. $15.95 each title

Titles: *Executive Fitness, Executive Nutrition and Diet,* and *Coping with Executive Stress.*

Business people will appreciate this easy-to-read series on wellness. Executive Health Examiners (EHE) was founded 20 years ago to make medical knowledge available to executives for their particular health needs and lifestyles. These volumes combine medical facts with EHE's experience in providing health care to the business community. Eating properly, exercising regularly, and coping with stress are habits that EHE editors encourage executives to acquire for good health maintenance.

Staying Healthy. A Bibliography of Health Promotion Materials. November 1984. U.S. Department of Health and Human Services. Single copies free.

Order this resource. Forty-two-page pamphlet is a guide to current information on health promotion and disease prevention topics. Describes materials, primarily booklets, fact sheets, films, posters, program guides, and reports. Full ordering information is given for each item. Comprehensive topics, but of special importance are health promotion, smoking, physical fitness and exercise, alcohol and alcoholism, and drugs and drug abuse. Updated frequently. Order directly from Superintendent of Documents, Washington, DC 20402.

Smoking

American Cancer Society. "Model Policy for Smoking in the Workplace." Available from local ACS units.

A sensible booklet that answers questions about how to go about respecting the rights of smokers and nonsmokers in drafting a smoking policy.

American Lung Association. "Taking Executive Action" and "Creating Your Company Policy." Available from local ALA affiliates.

Booklets that offer a useful employee survey form and guidelines for writing a smoking policy. Shows several versions of company policies and discusses the problems of ventilation, air purification, smoking in shared work spaces, and ongoing enforcement and conflict resolution.

Office on Smoking and Health, Department of Health and Human Services. *A Decision Maker's Guide to Reducing Smoking at the Worksite.* 1986.

Co-sponsored by the American Cancer Society, American Heart Association, and American Lung Association, this guide was prepared for the Office of Disease Prevention and Health Promotion and the Office on Smoking and Health. Must reading for any company. Available through local American Cancer Society units or from the Office of Smoking and Health, Park Building, 12420 Parklawn Dr., Rockville, MD.

Orleans, C. Tracy and John M. Pinney. *Nonsmoking in the Workplace: A Guide for Employers*. Washington, D.C.: Health Insurance Association of America, 1984. $50

The definitive guidebook on nonsmoking programs for employers. Available directly from the Public Relations Division of the Health Insurance Association of America, 1850 K St., N.W., Washington, DC 20006-2284.

Alcohol and Drug Abuse

Levy, Stephen J. *Managing the Drugs in Your Life: A Personal and Family Guide to the Responsible Use of Drugs, Alcohol, and Medicine*. New York: McGraw-Hill, 1983. $7.95

The author's theme is "drugs are not a drug issue—drugs are a people issue." Levy avoids a moralistic approach to the use of medications, street drugs, and other stimulants such as caffeine and nicotine. Chapter 3 covers drugs in the workplace and the cost of drug abuse to employers. Health promotion by employers is discussed.

Long, James. *Essential Guide to Prescription Drugs*. New York: Harper & Row, 1985. $10.95

Long provides consumers with information about safe prescription drug use. Highly recommended for the layperson in place of the Physicians Desk Reference, which is published by drug manufacturers. Revised regularly.

Milam, James Robert and Katherine Ketcham. *Under the Influence: A Guide to the Myths and Realities of Alcoholism*. Seattle: Madrona Publishers, 1981. $12.95

Nutrition

Brody, Jane. *Jane Brody's Nutrition Book: A Lifetime Guide to Good Eating for Better Health and Weight Control*. New York: Norton, 1981. $12.98; Bantam, 1982. $9.95

Written by the personal health columnist of *The New York Times*, this is an excellent book for the layperson. Twenty-seven chapters outline these six topics: what to eat, noncaloric nutrients, what to do about your weight, food for special lives, what's in your food, what to drink. Very readable presentation of the nutrition information needed to sustain good health.

Department of Health and Human Services, Office of Disease Prevention and

Health Promotion, the Society for Nutrition Education, and the American Dietetic Association. *A Decision Maker's Guide to Nutrition Programs at the Worksite*. Chicago: American Dietetic Association, 1986.

A collaborative effort by the government agency and the two leading nonprofit nutrition organizations. Helps decision makers decide from among a bewildering array of programs and providers. Filled with nutrition ideas that work. Order from the ADA, 430 N. Michigan Ave., Chicago, IL 60611.

"Nutrition at the Worksite," supplement to the *Journal of Nutrition Education*, April 1986. $15.75.

The most complete resource to date on nutrition at the worksite. Includes a provider/site directory of programs. Readers can contact these program directors in business and industry for suggestions or information. Order from the Society for Nutrition Education, 1736 Franklin St., Oakland, CA 94612.

Exercise

Cooper, Kenneth H. *The Aerobics Program for Total Well-Being: Exercise, Diet, Emotional Balance*. New York: M. Evans, 1982. $16.95; Bantam, 1983. $10.95

Cooper changed America's exercise programs with his aerobics books. This volume is an update of his research on nutrition and exercise physiology. For those concerned with total well-being, this is worthwhile reading. It includes examples of exercise and fitness categories. Bibliographies included.

Patton, Robert W., James M. Corry, Larry R. Gettman, Joleen Schovee Graf. *Implementing Health/Fitness Programs*. Champaign, Ill.: Human Kinetics Publishers, 1986.

A book for the fitness director or person who is responsible for setting up a corporate program. Lots of practical information on health risk appraisals, how-to information on programs and facilities.

Stress

Eliot, Robert S. and Dennis L. Breo. *Is It Worth Dying For? A Self-Assessment Program to Make Stress Work for You Not against You*. New York: Bantam, 1984. $15.95

A physician specializing in preventive cardiology and an editor with the American Medical Association have written a practical book to assess and correct the stressful aspects of modern life. They address the physical implications of stress and the management of the causes. Good bibliography.

Maddi, Salvatore R. and Suzanne C. Kobasa. *The Hardy Executive: Health under Stress.* Homewood, Ill.: Dow Jones-Irwin, 1984. $19.95

The authors studied middle- and upper-level executives and then developed their concept of "hardiness." Based on their research, this term may be translated as commitment, control, and challenge. This study would be useful for long-term planning in an organization rather than for reading by the individual employee.

Matteson, Michael T. and John M. Ivancevich. *Managing Job Stress and Health: The Intelligent Person's Guide.* New York: Free Press, 1982. $14.95

Authors say, "Stress is the cause—not the result—of physical and mental strain." They provide a practical readable book for anyone in any work environment.

Veniga, Robert L. and James P. Spradley. *The Work/Stress Connection: How to Cope with Job Burnout.* Boston: Little Brown, 1981. Ballantine, $3.50

This is one of the best books available about stress on the job. A must for any employer who wants to create a good working environment for the employees. This is also an excellent volume for the employee who is experiencing work stress or job burnout.

CHAPTER 4: WELLNESS COUNCILS CREATE A HEALTHY COMMUNITY

Healthy People: The Surgeon General's Report on Health Promotion and Disease Prevention. Office of Disease Prevention and Health Promotion, Department of Health and Human Services. 1979. $5

This landmark document sets forth some new priorities for the nation's health and calls for a renewed commitment to prevention. Order from the Superintendent of Documents, Washington, DC 20402.

Promoting Health/Preventing Disease: Objectives for the Nation. Public Health Service, Department of Health and Human Services. 1980. $5

A companion document to the national strategy set forth in *Healthy People*, identifies specific and measurable objectives for 15 priority areas, many which have direct implications for employers and the workplace. Order from the Superintendent of Documents, Washington, DC 20402.

CHAPTER 5: WELLNESS COUNCILS: HOW THEY WORK

The Wellness Council Approach to Wellness at the Worksite. Washington, D.C.: Health Insurance Association of America, 1986. $45 videocassette available in various formats

An 18-minute videocassette available for free preview showings to groups of chief executive officers and other business and civic leaders interested in forming a Wellness Council. Includes live-action scenes illustrating activities at Wellness Council sites around the country, narrated by Dr. Art Ulene and featuring an introductory statement by Dr. Otis R. Bowen, Secretary of the Department of Health and Human Services. Comes with a 52-page manual on operating a Wellness Council. Order from the Public Relations Division of the Health Insurance Association of America, 1850 K St., N.W., Washington, DC 20006-2284.

CHAPTER 6: THE INSURANCE INDUSTRY TAKES THE LEAD

The Health Insurance Association of America exclusively publishes and distributes several outstanding books, pamphlets, newsletters, and videocassettes on the topic of health promotion at the worksite. Some of them are described in other parts of this recommended reading section. HIAA also publishes a national newsletter, *Worksite Wellness Works*, four times a year. Subscriptions are currently available at no cost. For more information and to order, contact the Public Relations Division of the Health Insurance Association of America, 1850 K St., N.W., Washington, DC 20006-2284 or phone (202) 862-4122.

Wellness and the Bottom Line. Washington, D.C.: Health Insurance Association of America, 1984. $30 each in various videocassette formats

A 20-minute videocassette presentation of highlights from the HIAA video teleconference originating live via satellite from the Department of Health and Human Services in Washington, D.C., on March 13, 1984. Features President Ronald Reagan; Margaret Heckler, former secretary of HHS; Dr. Arthur Ulene, TV health commentator; Dr. Charles Berry, former chief medical director of the space program; James Ketelsen, CEO of Tenneco; and John Pearson, CEO of Northwestern National Life Ins. Co. Order directly from the Public Relations Division of the Health Insurance Association of America, 1850 K St., N.W., Washington, DC 20006-2284.

Wellness at the Worksite: The Time Is Now. Washington, D.C.: Health Insurance Association of America. $45 videocassette

A 15-minute video presentation filmed on location at corporate sites across the country. A supplementary booklet describes how to get started, what to do, and where to look for help. Order directly from the Public Relations Division of the Health Insurance Association of America, 1850 K St., N.W., Washington, DC 20006-2284.

Wellness at the School Worksite: A Manual. Washington, D.C.: Health Insurance Association of America, 1985. $6

To help school administrators, especially elementary, intermediate, and secondary school principals, to develop and implement wellness programs for all building staff—instructional, administrative, and support. Includes useful worksheets and checklists. Order directly from Public Relations Division of the Health Insurance Association of America, 1850 K St., N.W., Washington, DC 20006-2284.

CHAPTER 7: WELLNESS IN THE YEAR 2000

Arthur Andersen & Co. and the American College of Hospital Administrators. *Health Care in the 1990s: Trends and Strategies.* 1984. $65

This is a compact, readable report that is the result of a study to determine the future of the health care system to the year 1990. It highlights the trends and strategies that are shaping health care in the United States. The study provides sections on marketing of health care services, the payment for health care, and the types of care that the public can expect to receive in the future. This information should be useful to businesses as they plan health care services for their employees. Order from Compuletter, 5616 N. Western Ave., Chicago, IL 60659.

Bezold, Clement, Rick J. Carlson, and Jonathan C. Peck. *The Future of Work and Health.* Dover, Mass.: Auburn House, 1986.

Identifies the most important characteristics of work and workplaces over the next 25 years. Looks at changes in the work force, the aging baby boom generation, compression of morbidity, changing work values, new technologies. Introduces key trends shaping the future of health and health care with implications for promotion of health in the workplace.

ABOUT THE AUTHOR AND CONTRIBUTING AUTHORS

WILLIAM M. KIZER

William M. Kizer is chairman of the board of Central States Health & Life Company of Omaha. He has been a mover and a shaker in the wellness movement—some even accuse him of being ahead of his time—but he continues to make great personal and financial commitments to promote wellness in his own company and for the Wellness Council of the Midlands (WELCOM).

As one of the founders of WELCOM, Bill Kizer's vision that "the private sector, not government, is the logical provider of the information, skills, and motivation needed to enable people to make informed choices about the way they live" has caught on in cities everywhere. His personal regimen for wellness includes moderate swimming, watching his diet, and passing the torch of the family-controlled company his father started during the Depression to his son—one of nine Kizer children. That freed Bill to pursue his wellness vision with even more vigor and intensity, and this book is just one major project he has moved from the back burner to the fire. Over the past 10 years, Central States has enjoyed impressive growth, and if you ask Bill Kizer, he'll modestly tell you it's not because of his leadership but because Central States of Omaha is a well new workplace in which employees pursue personal and career goals in a supportive environment.

A product of Creighton University, Bill Kizer continues to support his community by serving on the Creighton board of directors. He is past chairman of the health education committee of the Health Insurance Association of America and serves on the boards of the Southern

California Water Company in Los Angeles and the American National Bank in Omaha.

In 1985 the Department of Health and Human Services gave special recognition to Bill Kizer for his role in the founding of the Wellness Council of the Midlands.

FRED W. SCHOTT

Fred W. Schott, vice-president of training and development at Central States Health & Life Company, is past president of WELCOM. He is a sought-after speaker and workshop leader whose commitment to addressing the role of corporate America in worksite wellness and the health and self-esteem of families is no secret. Fred Schott has had a distinguished career as a professional youth worker and administrator in inner-city areas of Louisville, Chicago, and Omaha. In his own consulting business, he became one of the most successful and highly respected private trainers in the area, working with nonprofit human service organizations and most of the area's top corporations.

Fred was awarded scholarships for his undergraduate education in Kentucky through the Jessie V. and W. Clement Stone Boys Clubs Foundation and completed his graduate work in urban studies at the University of Nebraska with scholarships from the Boys Clubs of America and Boys Clubs of Omaha.

In his own privately published book series, Healthy Families, Fred says the growing worksite wellness movement goes along with corporate concern for the quality of work and the quality of family life. In his books, which are filled with common-sense approaches to serious family concerns, Fred introduces readers to his three teenage daughters and his wife, Donna—his best friend.

HAROLD S. KAHLER, JR.

Harold S. Kahler, Jr., executive director of the Wellness Council of the Midlands, brings to this book his in-the-trenches experience in health promotion and as a consultant to hospitals and businesses in setting up wellness programs and selecting appropriate community services.

A graduate of Mansfield State College with a degree in education, Pacific School of Religion with a master's degree in systematic theology, and Iowa State University with a doctorate in higher education, Harold brings a diverse educational background to the field of wellness.

Harold was a wellness specialist at the University of Wisconsin— Stevens Point, where he coordinated the National Wellness Conference from 1978 to 1980. While at the university, he designed, implemented, and evaluated numerous wellness programs for students, faculty, staff, and local business. He also consulted with business throughout the state.

Before coming to WELCOM, Harold was a health educator at Iowa State University and a health promotion director in a hospital setting. Harold's career has been inspired by his quest for a healthy lifestyle. He has been a runner for 15 years and during that time has realized the importance of a well-balanced lifestyle that includes care and nurturing of the social, psychological, and spiritual aspects of his being as well as the physical.

Harold and his wife, Sheryl, share the same interests, zest, and enthusiasm that helps to create and maintain a wellness lifestyle.

INDEX

183